Biocontrol of Plant Diseases

Biocontrol of Plant Diseases

Eric August

Larsen & Keller
www.larsen-keller.com

Biocontrol of Plant Diseases
Eric August
ISBN: 978-1-64172-442-5 (Hardback)

⊟ Larsen & Keller

Published by Larsen and Keller Education,
5 Penn Plaza,
19th Floor,
New York, NY 10001, USA

Cataloging-in-Publication Data

Biocontrol of plant diseases / Eric August.
    p. cm.
Includes bibliographical references and index.
ISBN 978-1-64172-442-5
1. Phytopathogenic microorganisms--Biological control. 2. Plant diseases. 3. Phytopathogenic microorganisms--Control. 4. Diseased plants. I. August, Eric.
SB732.6 .B56 2020
632.3--dc23

For more information regarding Larsen and Keller Education and its products, please visit the publisher's website www.larsen-keller.com

Table of Contents

Preface

Biocontrol refers to a method of managing plant diseases and pests through other organisms. Diseases in plants are caused by pathogens and environmental conditions. Biocontrol deals with the control of such pathogens such as fungi, bacteria, viruses, protozoa and viriods. It depends on natural mechanisms like predation, parasitism and herbivory. The main strategies which are used within biocontrol are augmentation, importation and conservation. In augmentation, a large number of natural pathogens are administered for quick pest control. Importation is a method where a natural enemy of a pest is brought for controlling pests. Conservation relies on measures which are aimed at maintaining natural enemies by regular reestablishment. This book elucidates the concepts and innovative models around prospective developments with respect to biocontrol of plant diseases. It is compiled in such a manner, that it will provide in-depth knowledge about the theory and practice of this field. This textbook is appropriate for those seeking detailed information in this area.

A short introduction to every chapter is written below to provide an overview of the content of the book:

Chapter 1 - Biocontrol is a method through which pests are controlled. It depends on natural mechanisms such as predation or parasitism but also includes human intervention and management. Biological control agents include competitors, parasitoids and pathogens such as fungi and bacteria. This is an introductory chapter which will introduce briefly all the significant aspects of biocontrol and how it is used to deal with plant diseases.; **Chapter 2** - Plant disease can be described as an impairment of the normal condition of the plant that affects its vital functions. The organisms which can cause such infectious diseases are known as plant pathogens such as viruses, oomycetes, bacteria, viroids, fungi, nematodes, etc. The chapter closely examines the key concepts of plant diseases and pathogens to provide an extensive understanding of the subject.; **Chapter 3** - There are various fungal and bacterial diseases which plague fruit bearing trees and plants. A few examples of such diseases are fire blight, peach scab and brown rot. The topics elaborated in this chapter will help in gaining a better perspective about the diseases which affect different fruit crops as well as how they can be treated using biological control.; **Chapter 4** - Diseases of seed can affect seeds or seedlings and are generally responsible for thin stands and poor emergence. Such diseases cause damping-off, fading out or seedling blight. The topics elaborated in this chapter will help in gaining a better perspective about the different types of diseases of seed as well as their treatments.; **Chapter 5** - Rust refers to a number of plant diseases which are caused by pathogenic fungi. It commonly affects wheat and various ornamental crops like marigold and fuchsia. Powdered mildew is another fungal disease which affects a large number of crops such as wheat, barley, grapes and onions. The diverse applications of biocontrol in treating rust and powdered mildew have been thoroughly discussed in this chapter.; **Chapter 6** - Weeds are unwanted plants in human controlled settings such as lawns, farms and fields. Weed control is a vital method in agriculture that aims to stop the growth of weeds. There are various methods employed to control weeds. This chapter discusses in detail the theories and methodologies related to the biocontrol of weeds using microbes.

I extend my sincere thanks to the publisher for considering me worthy of this task. Finally, I thank my family for being a source of support and help.

Eric August

Biocontrol of Plant Diseases: An Introduction

Biocontrol is a method through which pests are controlled. It depends on natural mechanisms such as predation or parasitism but also includes human intervention and management. Biological control agents include competitors, parasitoids and pathogens such as fungi and bacteria. This is an introductory chapter which will introduce briefly all the significant aspects of biocontrol and how it is used to deal with plant diseases.

Biocontrol

Biological control is a method of restricting effects of harmful animals, pathogens and plants using other useful organisms, e.g. microorganisms, insects and plants that inhibit the harmful organisms.

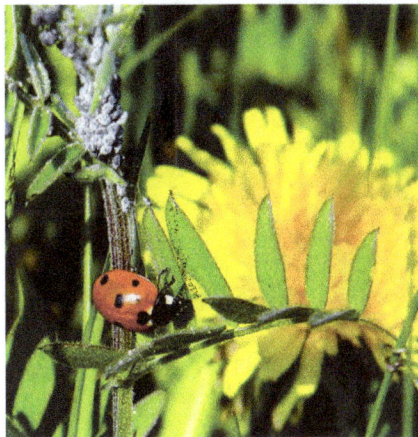

The method takes advantage of basic ecological interactions between organisms, such as predation, parasitism, pathogenicity and competition. Today, biological control is used primarily for controlling pests in crop cultivation.

Advantages of biological control are that no artificial substances are added, and that pathogens/ animals that develop resistance against biological control agents are rare. Biological control is an important component of integrated pest management.

Three Different Types of Biological Control

There are three basic types of biological control strategies:

1. Classical biological control means that you intentionally introduce a natural enemy or antagonist to a pest in a new region, with the intention of establishing and spreading the introduced organism in the new environment and fight the target pest.

2. Augmentative biological control - here, natural enemies are added, either to strengthen an already existing population or to temporary (not permanently) establish a population of natural enemies. An example of the latter case is insects or arachnids used in greenhouses.

3. Conservation biological control involves a deliberate modification of the environment to protect and promote natural enemies or antagonists to pests.

Using biological control methods is well established in commercial greenhouse cultivation. It also works well for hobby gardens.

One kind of biological control is to promote enemies to the pests that occur naturally in the environment. Ladybugs are good at eating aphids that damage the crops.

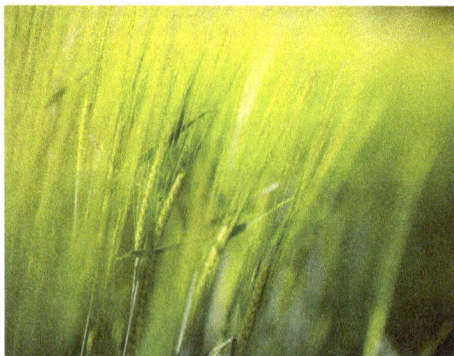

The biological pesticides Cedomon and Cerall have, since being introduced in the late 90s, replaced about 1.5 million liters of synthetic fungicides in the cultivation of cereals.

Uses of Biological Control

Biological control is a broad concept and can be used for the control of pests in very many different contexts.

Agriculture

Plant pests can be limited by means of natural enemies such as fungi, bacteria or arachnids. Biological control can be used to combat insects and fungal diseases as well as damages from nematodes. There are a variety of organisms and products that can be used in the cultivation of many different crops e.g. cereal seeds, oilseeds, potatoes, corn, peanuts and cotton.

Horticulture

Biological control works well in greenhouses as it is quite easy to control the environment to create good conditions for different organisms. It is important to quickly apply the beneficial organisms when a pest emerges. In this way, it is easier for the beneficial organism to control the pest. Today, biological methods are used in greenhouses for the control of eg. greenhouse spider mites, mildew, aphids, thrips, mealybugs and larvae of dark-winged fungus gnats, weevils and butterflies.

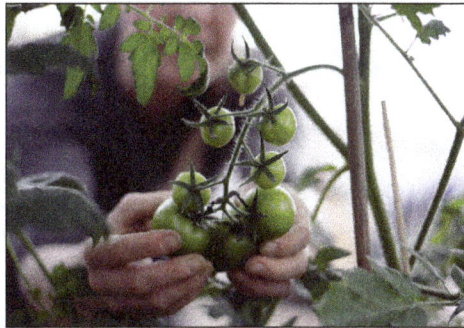

In a research project, Hanna Friberg from the Department of Forest Mycology and Plant Pathology, examines how biological control can be used to overcome corky root rot in organic tomato cultivation.

Forestry

In forestry, biological control is for example used to reduce attacks by Heterobasision annosum, a pest that causes losses in forestry of EUR 50-100 million annually. The biological pesticide is called Rotstop and prevents Heterobasidion from spreading and cause root rot in spruce and pine populations. Rotstop contains live spores of a fungus found naturally in our forest - Phlebiopsis gigantea. Phlebiopsis gigantea is a fungus that grows on dead wood and old timber rolls.

Examples of Organisms used in Biological Control

Bacteria

Field work in an area by the river Dalälven where Bacillus thuringiensis has been used against mosquito larvae.

Bacteria are used with great success as biological pesticides and can have many positive effects on the plants. The biological control products Cedomon and Cerall have, since their introduction in the late 1990s, replaced about 1.5 million liters of synthetic fungicides in the cultivation of cereals. Another well-known example is the bacterium Bacillus thuringiensis, which is used worldwide for larvae of insect pests.

Insects and Arachnids

In greenhouse cultivation, cultivated insect predators and parasitoids or predatory mites have been added for a long time in order to inhibit pests. In greenhouses you can create good conditions for the natural enemies, and biological control is often an effective measure if it is set up early. In open cultivation systems covering large areas, it is more difficult to use augmentative biological control. Instead, we can apply conservation biological control and promote enemies to the plant pests that naturally occur in the environment.

A crab spider catches a hoverfly.

Nematods

Nematodes are millimeter-long roundworms that live in the water film that surrounds soil particles. They can be used in biological control as several species can cause diseases in pests and other invertebrates. Some nematode species are currently used commercially for the control of dark-winged fungus gnats and snails. Nematodes can also fight pathogenic fungi. These fungi-eating nematodes have a stylet which they insert into fungal cells to eat the cellular contents, causing the fungal cells to die.

Fungi

Antagonistic fungi are important tools in biological control of plant pathogenic fungi. Many antagonistic fungi are found in the genus Trichoderma, but there are several other examples, some of which have been developed and marketed for biological control of plant diseases. These fungi are active against soil-borne pathogens, for example species within the genera Pythium, Fusarium and Rhizoctonia. Fungi can also be used above-ground, an example is Ampelomyces quisqualis used for mildew in horticulture. Yeast fungi have also been used to control mold damage during storage, such as Cryptococcus albidus and Candida oleophila against species within Botrytis and Penicillium.

At CBC, we investigate the fungus Clonostachys rosea, which has been shown to be effective against several plant diseases. Among other things, we examine how it best can be used against the fungus Fusarium graminearum which causes diseases on wheat and barley.

Principles of Biological Control

Because all insects and pests have some natural enemy, the enemies must be managed. Management of natural enemies is achieved through controlling importation, directing conservation, and monitoring growth. The goal of biological control is for the natural enemy introduced, to establish itself and continue to provide control without assistance from the gardener or farmer. Standards for biological control include:

1. One living organism is used to control another living organism.

2. Some control organisms require a limited host range and are therefore considered host specific.

3. Biological control agents affect the organisim either directly or indirectly.

   - Direct – kills the pest

   - Indirect – weakens the host so they can not reproduce at a normal rate

Three types of Applied Biocontrol that man can Influence:

Augmentation

Man can increase the native agents for control. Normally, there is a lack or absence of natural enemies occurring in the early pest season. Man can release the natural enemy early in the season to ensure that when the pests first appear natural enemies will not be scarce.

Classical Biological control

Man can introduce exotic biological control agents from their native home into the areas where exotic pests have established themselves and survived and multiplied due to the absence of natural enemies. This way, the control agents will reestablish equilibrium to keep the pest under control. This method is referred to as an old association.

Neoclassical biological control

Man can introduce an exotic biological control agent that previously did not have an association with the pest. The new biological control agent can establish itself and prey on the pest. This is referred to as a new association.

Natural Enemies

There are four types of natural enemies that can be used in biological control.

Parasitoid

A parasitoid is an organism that spends a significant portion of its life history attached to or within a single host organism which it ultimately kills (and often consumes) in the process. They are similar to typical parasites except in the fate of the host. In a typical parasitic relationship, the parasite and host live side by side without lethal damage to the host. The parasite takes enough nutrients to thrive without preventing the host from reproducing. In a parasitoid relationship, the host is killed before it can produce offspring.

This type of relationship seems to occur only in organisms that have fast reproduction rates (such as insects or mites). Parasitoids are also often closely coevolved with their hosts.

Insects Parasitoids

An insect parasitoid is an insect parasite that destroys its host. It has an undeveloped life stage. They are valuable as natural enemies because they develop on or within a single insect host and prevent any further development of the host after initial parasitization, eventually killing the host. This typically involves a host life stage which is immobile (e.g., an egg or pupa), and almost without exception they live inside the host. Insect parasitoids only damage a specific life stage of one or several related groups.

Characteristics of Insect parasitoids

- Specific in their choice of host.
- Smaller than their host.
- Only females search out hosts.
- Different parasitoid species attack different development stage of the host.
- Adults are free living, mobile and may be predatory.
- Undeveloped usually kill the host.

Usefulness as Biocontrol

An insect predator will instantly kill or immobilize their prey. A pest attacked by a parasitoid will die more gradually. The presence of parasitoids may not be obvious, even though they may be effective. In some cases, it is necessary to dissect or raise samples of pest insects to determine if any adult parasitoids appear. Because parasitoids are often more susceptible to chemical insecticides than their hosts, caution must be used.

Bathyplectes spp. are small, non-stinging wasps that are parasitoids of the alfalfa weevil, a serious pest of alfalfa in the Midwest and elsewhere. They were introduced to North America in 1911, from Italy by the U.S. Department of Agriculture as part of a biological control effort against the alfalfa weevil.

Predators

Insect predators are established in practically all agricultural and native habitats. The following are natural enemies of cabbage pests.

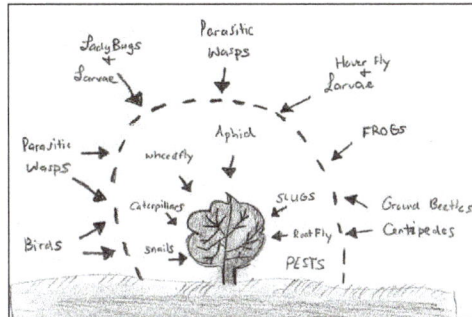

One of the benefits, of insect predators is that they are found all around and throughout plants, including below ground, as well as, trees and shrubs. Some predators are specific in their choice of prey, while others are not and they need a wide range of prey. Because they destroy their prey quickly, their successes are easily recognized. However, specific predators may not be identified due to their mobility. Once they feed, they move on. Major characteristics of arthropod predators:

- The adults and young are usually generalists.

- Normally larger than their prey.

- They destroy or consume many prey.

- They will attack both immature and adult prey.

Examples

Ladybugs, and in particular their larvae which are active between May and July, are voracious predators of aphids such as greenfly and blackfly, and will also consume mites, scale insects and small caterpillars.

Hoverflies are another very welcome garden predator. Resembling slightly darker bees or wasps, they have characteristic hovering, darting flight patterns. There are over 100 species of hoverfly whose larvae principally feed upon greenfly, one larva devouring up to fifty a day, or 1000 in its lifetime. They also eat fruit tree spider mites and small caterpillars. Adults feed on nectar and pollen, which they require for egg production.

Aphids, also known as greenfly/blackfly

Wax scales on a lemon tree branch

Female Syrphid (Hoverfly) Fly

Phacelia calthifolia

Hoverflies can be encouraged by growing attractant flowers such as the poached egg plant marigolds or phacelia throughout the growing season.

Dragonflies are important predators of mosquitoes, both in the water, where the dragonfly nyads eat mosquito larvae, and in the air, where adult dragonflies capture and eat adult mosquitoes. Community-wide mosquito control programs that spray adult mosquitoes also kill dragonflies, thus removing an important biocontrol agent and can actually increase mosquito populations in the long term. Other useful garden predators include lacewings, rove and ground beetles, aphid midge, centipedes, predatory mites, as well as mega fauna such as frogs, toads, hedgehogs, slowworms and birds.

Usefulness as Biocontrol

Although most beneficial predators consume large amounts of pest insects during their life span some predators are more effective at controlling pests than others. Some species play an important role in the containment of pests and others provide good late season control. Some play a minor roll by themselves but their contribution greatly influences overall pest mortality.

A good example of the potential number and diversity of predators in a crop comes from an Agricultural survey on cotton crops in Arkansas. More than 600 species of predators in 45 different families of insects and 23 families of spiders and mites have been documented, in Arkansas cotton. In the northeastern United States, eighteen species of predatory insects (not including spiders and mites) have been found in potatoes. Within a single acre, there may be thousands of predators in addition to many parasitoids. Although the effect of any one species of natural enemy may be minor, the combined influence of predators, parasitoids, and insect pathogens can be significant.

Pathogens

A pathogen or infectious agent is a biological agent that causes disease or illness to its host. Disease causing organisms such as bacteria, viruses, protozoa and fungi can infect insects and mites. These natural occurring organisms can multiply to cause disease outbreaks or epizootics on pest populations. Under the right conditions (high humidity, high pest numbers), an outbreak of epizootics can eliminate an insect population. Diseases are an essential and normal control for some insect pests. Microbial insecticides, biorational, or bio-insectices are pathogens that have been mass produced and are accessible in commercial formulations for use in standard spray equipment.

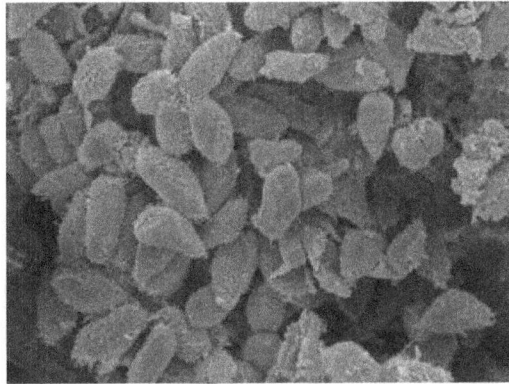

While some microbial insecticides are still experimental, others have been available for use for several years. Gardeners and commercial growers have widely used formulations of the bacterium, Bacillus thuringiensis. One of the benefits of using microbial products is that they do not directly affect beneficial insects and none are hazardous to wildlife or humans. Most insect pathogens target certain groups of insects at certain life stages.

Chemical insecticides may bring about quicker results as microbial insecticides can take longer to destroy or weaken a target pest, which may limit their use for crops that can maintain some insect damage. In order for the use of microbial insecticides to be most effective they must be applied at the correct life stage of the pest. Because the use of microbial insecticides is compatible with the use of predators and parasitoids this aids in spreading some pathogens through the pest population. Although microbial insecticides are considered non-toxic to humans, safety precautions must be followed to minimize any exposure.

Characteristics of Insect Pathogens

- Use will result in killing, reducing reproduction, slowing growth, or shortening the life of pests.

- Specific to target species or specific life stages.

- Effectiveness is often dependent upon other factors, such as host abundance or environmental conditions.

- Degree of control may not be predictable.

- Normally slow acting; adequate control may not take place for several days or longer Usefulness as Bio-control.

Due to the fact that effectiveness is often dependent upon other factors and the proper deployment appears to be crucial, disease control in the field is likely to be less successful then in the laboratory where ideal conditions are in place. Continued research needs to be carried out, as this may be another effective method for control.

Weed-feeders

When new exotic plant species are introduced to new locations through out the world, problems often arise. The new species may spread rapidly because there are no effective natural enemies in the new location. It is estimated that 50% to 75% of the problem weeds in the U. S. arrived from other areas either accidentally or intentionally. An example would be purple loosestrife, a semi-aquatic herbaceous plant belonging to the loosestrife family. Purple loosestrife is native to the wetlands of Eurasia and was introduced to America in the late nineteenth century. In North America because of the lack of natural enemies, it is a nuisance choking water ways, and crowding out native North American species.

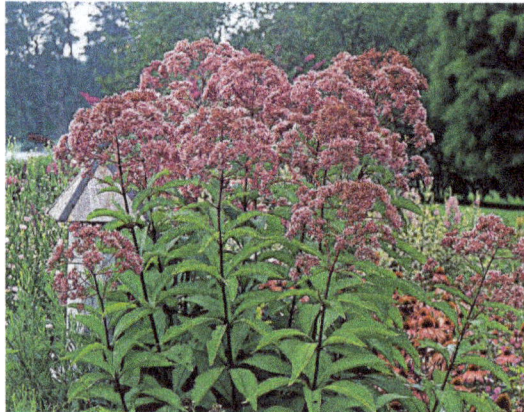

Types of Weed Feeders

Fungal products, insects, plant pathogens, nematodes and fish can be effective in controlling pest weeds. Insects feed on seeds, flower leaves, stems, roots or a combination of all of these. Insects can also pass on plant pathogens to infest the plants.

It is important to determine if the weed feeder will feed exclusively on the pest species before introducing it. Effectiveness of control, presence of a favorable host plant, harmonization with the natural enemy's life cycle, productiveness, heartiness, and similarities in climate and ecology should also be considerations.

The fungus, Colletotrichum gloeosporioide has been proven effective in controlling northern joint vetch, a plant pest found in rice and soybean crops. Colletotrichum gloeosporioide causes plants to wilt and the crown tissues to decay. This fungus survives on infected plants and debris eventually, eliminating jointvetch from the fields.

Other Control Success in North America

- Nodding Thistle (Canada, Kansas U.S.).

- Ragwort (British Columbia, Canada, California and Oregon).
- Alligator Weed (Florida, Louisiana and Texas).

Characteristics of Weed-feeding Natural Enemies:

- Specific to One Plant Species.
- Negative impact on plant individuals and the population dynamics of the target weed.
- Fertile.
- Thrive and become widespread in all habitats and climates that the pest weed occupies.
- Good colonizers.

The introduction of weed-feeding natural enemies in North America have had mixed success. In some cases, they have been extremely effective with a 99% reduction of the pest species. In other cases, it has been complete failure, as the introduced species was not able to establish itself in the new location. Some of the failure may be attributed to predators, parasitoids of the newly introduced species limiting its ability to control the pest plant.

Required Steps for Biological Control

The USDA, APHIS (U.S. Department of Agriculture's Animal and Plant Health Inspection Services) is responsible for providing leadership in ensuring the health and care of animals and plants. PPQ (Plant Protection and Quarantine) which is part of the APHIS protects America's agricultural and ecological resources through their pest detection program. The pest detection program provides a continuum of checks from offshore pre-clearance programs through port inspections to surveys in rural and urban sites across the U.S. Because pests, weeds and diseases are potential agent of bioterrorism, a new aspect of the Department is biological terrorism. The APHIS grants permission for the release of any biological control agent or natural enemy. In individual states, the State Entomologist reviews requests prior to the release of any biological control agents. Research scientists conduct host specificity studies and submit the study results to federal and state agencies for their review. Once approval is obtained, exotic biological control agents may be released.

The required steps include the following:

1. Determine if the pest is appropriate for biological control.
2. Selection of suitable and effective natural enemies.
3. Conduct safety tests. Make sure that the potential control agents only attack the target pests, and not turn into pests themselves.
4. Develop techniques for raising the control agents to provide sufficient distribution.
5. Conduct field establishment studies prior to the release.
6. Evaluate surveys to assess the effectiveness of the biological control agent in controlling the pest.

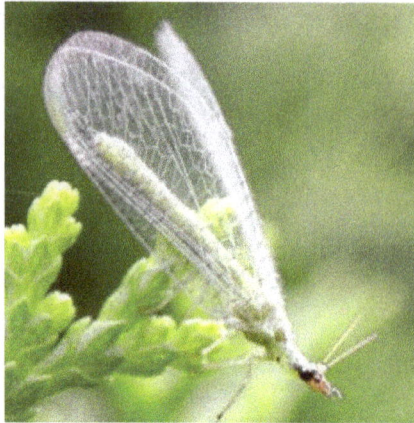

Predators of Pests

Green Lacewings

Green lacewings are insects in the very large family Chrysopidae of the order Neuropetra. There are approximately 1300 species. Lacewings are widespread insects; the genus Chrysoperla is very common in North America. Their larvae are voracious predators, attacking most insects of suitable size, especially soft-bodied ones (aphids, caterpillars and other insect larvae, insect eggs). Adults use substrate vibrations as a form of communication, especially during courtship; species which are nearly identical morphologically may sometimes be separated more easily based on their mating signals.

In several countries, millions of lacewings are reared for sale as biological control agents of insect and mite pests in agriculture and gardens. They are distributed as eggs, since they are highly aggressive and cannibalistic. The eggs hatch in the field, originating the predatory larvae. Their performance is variable; thus, there is a lot of interest on further improvement of the use of lacewings as biological pest control.

Insidious Flower Bug (Orius)

Orius is a predator of small insects and mites and is found on many agricultural crops including: cotton, peanuts, alfalfa, corn, pea and strawberry. It is a successful biological control agent in greenhouses. The immature stages (nymphs) and adults both feed on a variety of small prey. It is considered an excellent predator of the eggs and new larvae of the bollworm and the spotted tobacco aphid. It is believed that thrips and mites are the most basic part of its diet. It is also know the prey on corn leaf aphids potato aphids and potato leafhopper nymphs.

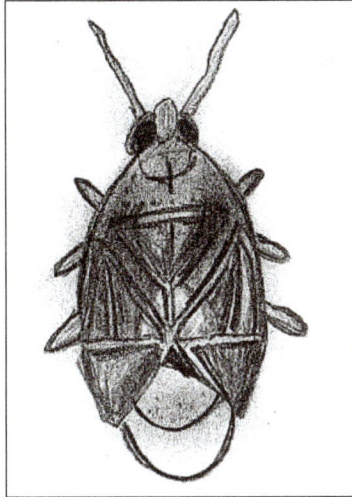

Bigeyed Bugs (Geocorus)

Bigeyed bugs are the most abundant and considered the most important predaceous insects in many corps throughout the U.S. With approximately 19 species and rarely causing economic damage they offer a great benefit to biological pest control cropping systems in the U.S. They are one of the most valuable natural enemies for cotton. They feed on eggs and small larva of bollworm, pink bollworm, tobacco bud worm, and on the eggs and nymphs of plant bugs. They also feed on all life stages of whiteflies, mites and aphids.

Harvestmen Spiders also known as Daddy Long Legs or Harvest spiders feed on many soft bodied arthropods in corn,alfalfa, small grains, potatoes, strawberries and apple crops.

Stink bugs are names for the strong odor they emit when disturbed. It is found throughout the U. S. and feeds on immature insects including: larvae of Mexican bean beetle, European corn borer, diamondback moth, corn earworm, beet armyworm, fall armyworm, cabbage looper, imported cabbageworm, colorado potato beetle, velvetbean caterpillar, and flea beetles.

Persimillis – A mite used in strawberry fields, greenhouse and warm humid habitats in which spider mites are problematic. Every fall and winter, millions of these mites are released in to the California strawberry fields and are an integral part of an IPM program of mite control.

Californicus – Feeds on spider mites, cyclamen and broad mites. It is able to tolerate higher temperatures and lower humidities than the perismillis mite. European growers use them in greenhouses to protect peppers and other crops from spider mites.

Helveolus – Is used to control the Persea mite in California avocado groves.

Parasites of Pests

Trichogramma

The wasps of genus Trichogramma are some of the most widely-studied agents of biological control in the field of entomology. Trichogramma wasps are tiny Hymenopteran insects, measuring 1 millimeter in length or less, that parasitize the eggs of many types of agricultural pest insects. They are easy to rear and release in fields suffering from pest outbreaks.

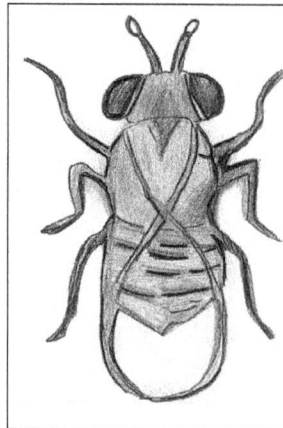

There are over 230 species of Trichogramma, and most are so similar that advanced expertise is required to tell them apart. Genetic studies are ongoing. The wasps are currently used to control at least 28 species of insect pest, including the cotton bollworm, codling moth and corn borer. Female wasps inject their own eggs into the egg of the pest, and her larva consumes the embryo and other contents of the egg.

Encarsia

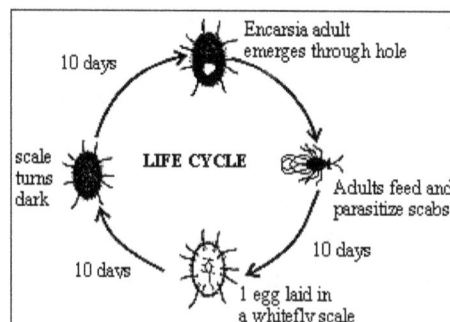

Encarsia inaron is a tiny parasitoid, stingless wasp first collected in Italy and Israel and introduced into California in 1989 to control the ash whitefly that was collected in Italy and Israel and introduced into California in 1989 for the control of the ash whitefly. The eggs of the Encarsia develop inside the whitefly host.

Approaches to Biological Control

There are three general approaches to biological control; importation, augmentation and conservation of natural enemies. Each of these techniques can be used either alone or in combination in a biological control program.

Importation

Importation of natural enemies, sometimes referred to as classical biological control, is used when a pest of exotic origin is the target of the biocontrol program. Pests are constantly being imported into countries where they are not native, either accidentally, or in some cases, intentionally. Many of these introductions do not result in establishment or if they do, the organism may not become pests. However, it is not uncommon for some of these introduced organisms to become pests, due to a lack of natural enemies to suppress their populations. In these cases, importation of natural enemies can be highly effective.

Once the country of origin of the pest is determined, exploration in the native region can be conducted to search for promising natural enemies. If such enemies are identified, they may be evaluated for potential impact on the pest organism in the native country or alternatively imported into the new country for further study. Natural enemies are imported into the US only under permit by the US Department of Agriculture. They must first be placed in quarantine for one or more generations to be sure that no undesirable species are accidentally imported (diseases, hyperparasitoids etc.). Additional permits are required for interstate shipment and field release.

Adult Bathyplectes anurus, a parasitoid of alfalfa weevil larvae.

Biological control of the alfalfa weevil, Hypera postica (Gyllenhall) is a example of a successful program using importation of natural enemies. The alfalfa weevil, a native of Europe, was originally detected in the US in Utah in 1904. A second introduction was detected on the East coast in 1951. By 1970, the weevil had spread to all 48 contiguous states and become a serious pest of alfalfa. Some importation's of natural enemies began as early as 1911, however, a major program aimed at biological control of the weevil was initiated in 1957. In this program, USDA ARS personnel conducted foreign exploration in Europe resulting in the eventual importation of 12 parasitoid species. Six of these species became established and are credited with contributing to the reduction in the weevil's pest status in the eastern US.

Augmentation

Augmentation is the direct manipulation of natural enemies to increase their effectiveness. This can be accomplished by one, or both, of two general methods: mass production and periodic colonization; or genetic enhancement of natural enemies. The most commonly used of these approaches is the first, in which natural enemies are produced in insectaries, then released either inoculatively or inundatively. For example, in areas where a particular natural enemy cannot overwinter, an inoculative release each spring may allow the population to establish and adequately control a pest. Inundative releases involve the release of large numbers of a natural enemy such that their population completely overwhelms that of the pest. Augmentation is used where populations of a natural enemy are not present or cannot respond quickly enough to the pest population. Therefore, augmentation usually does not provide permanent suppression of pests, as may occur with importation or conservation methods.

An example of the inoculative release method is the use of the parasitoid wasp, Encarsia formosa Gahan, to suppress populations of the greenhouse whitefly, Trialeurodes vaporariorum. The greenhouse whitefly is a ubiquitous pest of vegetable and floriculture crops that is notoriously difficult to manage, even with pesticides. Releases of relatively low densities (typically 0.25 to 2 per plant, depending on the crop) of E. formosa immediately after the first whiteflies are detected on a greenhouse crop can effectively prevent populations from developing to damaging levels. However, releases should be made within the context of an integrated crop management program that takes into account the low tolerance of the parasitoids to pesticides.

Conservation

In any biological control effort, conservation of natural enemies is a critical component. This involves identifying the factor(s) which may limit the effectiveness of a particular natural enemy and modifying them to increase the effectiveness of the beneficial species. In general, conservation of natural enemies involves either, reducing factors which interfere with natural enemies or providing resources that natural enemies need in their environment.

Many factors can interfere with the effectiveness of a natural enemy. Pesticide applications may directly kill natural enemies or have indirect effects through reduction in the numbers or availability of hosts. Various cultural practices such as tillage or burning of crop debris can kill natural enemies or make the crop habitat unsuitable. In orchards, repeated tillage may create dust deposits on leaves, killing small predators and parasites and causing increases in certain insect and mite pests. In one study, periodic washing of citrus tree foliage resulted in increased biological control of California red scale, Aonidiella aurantii due to increased parasitoid efficiency. Finally, host plant effects such as chemical defenses which are harmful to natural enemies but to which the pest is adapted, can reduce the effectiveness of biological control. Some pests are able to sequester toxic components of their host plant and use them as defense against their own enemies. In other cases, physical characteristics of the host plant such as leaf hairiness, may reduce the ability of the natural enemy to find and attack hosts.

Ensuring that the ecological requirements of the natural enemy are met in the cropping environment is the other major means of conserving natural enemies. To be effective, natural enemies may need access to; alternate hosts, adult food resources, overwintering habitats, constant food supply, and appropriate microclimates. In a classic example, Doutt and Nakata determined that Anagrus epos Girault, principal parasitoid of the grape leafhopper, Erythroneura elegantula Osborne in California grape vineyards required an alternate host for overwintering. This host, another leafhopper, only overwintered on blackberry foliage in riparian areas, often quite distant from the vineyards. Vineyards close to natural blackberry stands experienced earlier colonization by the parasitoid in the spring and better biological control. Wilson et al. found that French prune trees which harbor another overwintering host, could be planted upwind of vineyards and effectively conserve Anagrus epos.

Current Applications of Biological Control

Biological control is an exciting science because it constantly incorporates new knowledge and techniques.

Modern Approaches in Augmentation of Natural Enemies

Because most augmentation involves mass-production and periodic colonization of natural enemies, this type of biological control has lent itself to commercial development. There are hundreds of biological control products available commercially for dozens of pest invertebrates, vertebrates, weeds, and plant pathogens.

The practice of augmentation differs from importation and conservation in that making permanent changes in a agroecosystem to improve biological control is not the primary goal. Rather, augmentation generally seeks to adapt natural enemies to fit into existing production systems. For example, cultures of the predatory mite, Metaseiulus occidentalis (Nesbitt) were laboratory-selected for resistance to pesticides commonly used in an integrated mite management program in California almond orchards. This program has saved growers $24 to $44 per acre per year in reduced pesticide use and yield loss. Genetic improvement of several predators and parasitoids has been accomplished with traditional selection methods, and appears possible with recombinant DNA technology.

An excellent example of an augmentative practice than has been successfully adapted to a wide variety of agricultural systems is the inundative release of Trichogramma wasps. These minute endoparasitoids of insect eggs are released in crops or forests in large numbers (up to several million/ha) timed to the presence of pest eggs. Trichogramma are the most widely augmented species of natural enemy, having been mass-produced and field released for almost 70 years in biological control efforts. Worldwide, over 32 million ha of agricultural crops and forests are treated annually with Trichogramma spp. in 19 countries, mostly in China and republics of the former Soviet Union.

In China, agricultural production and pest management systems capitalize on low labor costs, and generally follow highly innovative yet technologically simple processes. For example, Trichogramma spp. that arc inundatively released to suppress sugarcane borer, Chilo spp., populations in sugarcane are protected from rain and predators inside emergence packets. Insectary-reared parasitized eggs are wrapped in sections of leaves which are then slipped by hand over blades of sugarcane. Most Trichogramma production in China takes place in facilities producing material for a localized area. These facilities range from open air insectaries to mechanized facilities that are leading the world in development of artificial host eggs.

One of the barriers to wider implementation of biological control in western agriculture has been socio-economics. In current large-scale production agricultural systems, a premium is placed on efficiency and economy of scale. Entire support industries have developed around the application of agrichemicals, including application equipment manufacturing, distribution and sales, as well as application services. In order for biological control products to not be at odds with these industries, and to compete strongly with pesticides, they should have many of the same characteristics. Ideally, they should be as effective as pesticides, have residual activity, be easy to use, and they should have the capacity to be applied quickly on a large scale with conventional application equipment.

In Western Europe, almost two decades of intensive research resulted in the commercial marketing of three products utilizing the European native, Trichogramma brassicae Bezdenko, to suppress the European corn borer, Ostrina nubilalis Hübner, in corn fields. These products are annually applied to approximately 7,000 ha in each of Switzerland and Germany, 150 ha in Austria, and 15,000 ha in France. All three products are based on manufactured plastic or paper packets designed to provide protection for the wasps against weather extremes and predation until emergence in the field.

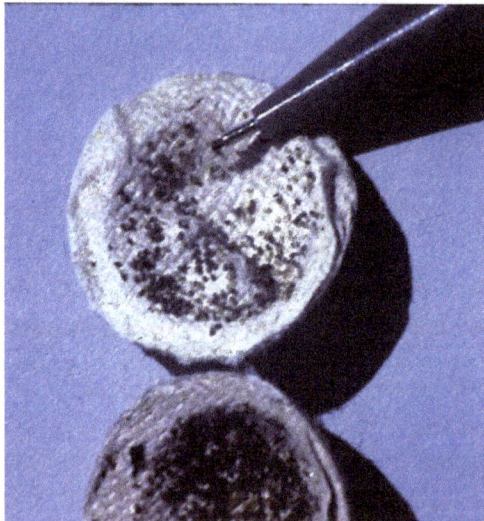

A Trichocap opened to show eggs parasitized by Trichogramma brassicae.
Upon emergence wasps exit through holes in capsule wall.

As in the Chinese example above, European Trichogramma products are for the most part applied to crop fields by hand. One exception is the product called, Trichocaps which can be broadcast either by hand or by aircraft using conventional application equipment. Trichocaps packets are actually hollow walnut-shaped cardboard capsules (2 cm. diam.) that each contain approximately 500 parasitized Mediterranean flour moth, Ephestia kuehniella Zwolfer, eggs. Developing Trichogramma inside capsules are induced into an overwintering (diapause) state in the insectary, then stored in refrigerated conditions for up to nine months without loss of quality. This system allows for production of product during winter months, then distribution to growers when needed in the summer.

Once removed from cold storage, Trichogramma inside the capsules will begin development and begin emergence approximately 100 Celsius degree days later. This 'reactivation' process can be manipulated so that capsules containing Trichogramma at different developmental stages can be applied to fields at the same time, extending the emergence period of parasitoids and increasing the 'residual' activity of a single application to approximately one week. Planning and preparation of the product for application is done by the company so that growers are only responsible for applying the product to crop fields.

Cooperative research over the last 5 years has resulted in successful commercial-scale pilot testing of this method in North America on seed corn and field corn production systems. This strategy now has the potential for immediate commercial implementation in North America.

Landscape Ecology and the Conservation of Natural Enemies

The study of disturbance and its effects on community dynamics and the emergence of the discipline of landscape ecology are impacting the way we think about the conservation of natural enemies. Over the past 15 years, ecologists have come to recognize the central role that disturbance plays in the structuring of ecological communities. While the most highly disturbed terrestrial ecosystems may have one disturbance event every several years (e.g. fire in grasslands), many agricultural ecosystems experience multiple events per growing season (plowing, planting, nutrient and pesticide applications, cultivation and harvest). From an ecological point of view, the outcomes are predictable. Highly disturbed systems exhibit decreased species diversity and shortened food chains, resulting in the few well adapted species (i.e. pests) having few natural enemies to suppress their populations. This requires that additional disturbance events be initiated (i.e. pesticide applications) which, while controlling the initial negative symptom, may precipitate its reoccurrence.

Current systems of crop production also shape the physical structure of our agricultural landscapes. With increased reliance on mechanization and pesticides, diversity in farmlands has rapidly disappeared and the impacts on natural enemies are only now beginning to be understood. In general, increased habitat fragmentation, isolation and decreased landscape structural complexity tend to destabilize the biotic interactions which serve to regulate natural ecosystems.

The goal of an ecological approach to conservation biological control is to modify the intensity and frequency of disturbance to the point where natural enemies can function effectively. This will need to occur at field, farm and larger landscape-levels. Within fields, modification of tillage intensity and frequency (reduced tillage or no-tillage) can leave more plant residue on the soil surface and have a positive impact on predators (ground beetles and spiders). Intercropping can also modify the microclimate of crop fields making them more favorable for parasitoids.

Female Eriborus terebrans a larval endoparasitoid of the European corn borer.

At the farm level, the presence and distribution of non-crop habitats can frequently be critical to natural enemy survival. Eriborus terebrans (Gravenhorst) is a wasp which parasitizes European corn borer larvae. Female Eriborus require moderate temperatures (<90° F) and a source of sugar (nectar of flowering plants or aphid honeydew). Neither of these conditions is met in a conventionally managed corn field. Therefore, wasps seek more sheltered locations in wooded fencerows and woodlots where they find reduced temperatures, higher relative humidity and abundant sources of adult food. European corn borer larvae in corn field edges near these types of habitats are parasitized at two to three times the rate of those in field interiors (up to 40%). Current research is examining the potential of modifying corn production systems by creation of natural enemy resource habitats to provide critical resources and increase natural control of European corn borers. Intercrops, strip crops, as well as modification of grass waterways, shelterbelts, buffer and riparian zones are promising techniques.

Typical field border with flowering plants important in providing pollen, nectar, alternate hosts and refuges for natural enemies of pests in agricultural landscapes.

Finally, at the landscape-level, the physical structure of agricultural production systems can also influence pest and natural enemy diversity and abundance. In a study contrasting simple versus mosaic landscapes, Ryszkowski et al. concluded that natural enemies are more dependent on refuge habitats than are pests and the greater abundance of these refuges in the mosaic landscapes resulted in their higher diversity, abundance and ability to respond to prey numbers. Marino and Landis (in press) examined parasitism of true armyworm, Pseudaletia unipuncta (Haworth), in structurally-complex versus simple agricultural landscapes. Overall parasitism in the complex sites was more than three times higher than in the simple sites (13.1% versus 3.4%). Differences were largely attributable to one wasp species, the braconid, Meterous communis (Cresson) which was far more abundant in complex sites. They hypothesized that abundance and proximity of preferred habitats for alternate hosts of M. communis may account for the observed differences.

In the past, conservation was typically attempted one species at a time, concentrating on meeting the needs of what was deemed the most important natural enemy in a particular system. While this will continue to be an enormously useful approach, it now seems possible that basic ecological theory could inform the design and management of landscapes to conserve and enhance the effectiveness of entire communities of natural enemies.

Interactions Between Plants and Beneficial Microbes

Throughout their lifecycle, plants and pathogens interact with a wide variety of organisms. These interactions can significantly affect plant health in various ways.In order to understand the mechanisms of biological control, it is helpful to appreciate the different ways that organisms interact. In order to interact, organisms must have some form of direct or indirect contact. The types of interactions between plants and microorganisms have been referred to as mutualism, protocooperation, commensalisms, neutralism, competition, amensalism, parasitism and predation. While the terminology has been developed for macroecology, examples of all of these types of interactions can be found in the natural world at both the macroscopic and microscopic level. And, because the development of plant diseases involves both plants and microbes, the interactions that lead to biological control take place at multiple levels of scale.

From the plant's point of view, biological control may be considered a positive result arising from different specific and non-specific interactions.We can begin to classify and functionally delineate the diverse components of ecosystems that contribute to biological control. Mutualism is an association among several species where all of them are benefited from this association. Sometimes, it can be an obligatory relation involving close physical and biochemical contact between two organisms, such as those between plants and mycorrhizal fungi. However, they are generally facultative and opportunistic.

For example, Rhizobium bacteria reproduce either in the soil or, to a much greater degree, through their mutualistic association with legume plants. These types of mutualism can contribute to biological control, by providing plant with improved nutrition and/or by stimulating host defense mechanism and ability. Many of the microorganisms isolated and classified as biocontrol agents (BCA) can be considered facultative mutualists, because host and disease suppression by them will vary depending on the prevailing environmental conditions.

Commensalism is also a symbiotic interaction between two living organisms, where one organism benefits and the other is neither harmed nor benefited. Most plant-associated microorganisms are assumed to be commensals with regards to the host plant, because their presence, individually or in total, rarely results in positive or negative consequences to the plant. While the presence of these microorganisms may present a variety of challenges to an infecting pathogen, their absence decreases pathogen infection or disease severity and is indicative of commensal interactions.

Biological interactions in which the population density of one species has absolutely no effect on the other are called neutralism. Related to biological control, an inability to associate the population dynamics of pathogen with that of another organism would indicate neutralism. In contrast, antagonism between organisms results in a negative outcome for one or both. Competition within and between species caused a decreased growth, activity, and/or fecundity of the interacting organisms. Biocontrol can occur when non-pathogens compete with pathogens for nutrients and sites in host plant. Direct interactions that benefit one population at the expense of another also affect our understanding of biological control.

Parasitism is also a symbiotic relation in which two organisms coexist over a prolonged period of time. In this type of interaction, one organism, usually the physically smaller (parasite) benefits and the other (host) is harmed. The activities of various hyperparasites, for example those agents that parasitize plant pathogens, can result in biocontrol. Another interesting contribution to biocontrol is when host infection and parasitism by relatively avirulent pathogens may lead to biocontrol of more virulent pathogens through the stimulation of host defense systems. Finally predation refers to the hunting and killing of one organism by another for consumption and sustenance. While the term predator typically refers to animals that feed at higher trophic levels in the macroscopic world, it has also been applied to the actions of microorganisms such as protists and mesofauna, e.g. fungal feeding nematodes and microarthropods, that consume pathogen biomass for sustenance.

Biological control can result in various forms of these types of interactions, depending on the environmental conditions within which they occur. Significant biological control, as was described above, generally arises from manipulating mutualisms between microorganisms and their plant hosts or from manipulating antagonisms between microbes and pathogens.

Mechanisms of Biological Control

Since biological control is a result of many different types of interactions among microorganisms, scientists have concentrated on characterization of mechanisms occurring in different experimental situations. In all cases, pathogens are antagonized by the presence and activities of other microorganisms that they encounter.

Direct antagonism results from physical contact and/or a high-degree of selectivity for the pathogen by the mechanism(s) expressed by the biocontrol active microorganisms. In this type of interaction, Hyperparasitism by obligate parasites of a plant pathogen would be considered the most direct type of mechanism because the activities of no other organism would be required to exert a suppressive effect. In contrast, indirect antagonism is resulted from the activities that do not involve targeting a pathogen by a biocontrol active microorganism. Improvement and stimulation of plant host defense mechanism by non-pathogenic microorganisms is the most indirect form of antagonism. While many studies have concentrated on the establishment of the importance of specific mechanisms of biocontrol to particular pathosystems, all of the mechanisms described below are likely to be operating to some extent in all natural and managed ecosystems.

The most effective biocontrol active microorganisms studied appear to antagonize plant pathogens employing several modes of actions. For example, pseudomonads known to produce the antibiotic 2, 4-diacetylphloroglucinol (DAPG) may also induce host defenses. Additionally, DAPG-producers bacterial antagonists can aggressively colonize roots, a trait that might further contribute to their ability to suppress pathogen activity in the rhizosphere of plant through competition for organic nutrients. However, the most important modes of actions of biocontrol active microorganisms are as follows:

1. Mycoparasitism: In Hyperparasitism, the pathogen is directly attacked by a specific biocontrol agent (BCA) that kills it or its propagules. Four major groups of hyperparasites have generally been identified which include hypoviruses, facultative parasites, obligate bacterial pathogens and predators. An example of hypoparasites is the virus that infects Cryphonectria parasitica, the fungal causal agent of chestnut blight, which causes hypovirulence, a reduction in pathogenicity of the pathogen. This phenomenon has resulted in the control of chestnut blight in many places. However, the interaction of virus, fungus, tree and environment determines the success or failure of hypovirulence.

In addition to hypoviruses several fungal hypoparasites have also been identified including those that attack sclerotia (e.g., Coniothyrium minitans) or others that attack fungal hyphae (e.g. Pythium oligandrum). In some cases, a single fungal pathogen can be attacked by multiple hyperparasites. For example, Acremonium alternatum, Acrodontium crateriforme, Ampelomyces quisqualis, Cladosporium oxysporum and Gliocladium virens are just a few of the fungi that have the capacity to parasitize powdery mildew pathogens.

In contrast to hyperparasitism, microbial predation is more general, non-specific and generally provides less predictable levels of disease control. Some biocontrol agents exhibit predatory behavior under nutrient-limited conditions. Such as Trichoderma, a fungal antagonist that produces a range of enzymes that are directed against cell walls of pathogenic fungi. However, when fresh bark is used in composts, Trichoderma sp. does not directly attack the plant pathogen, Rhizoctonia solani. But, in decomposing bark, the concentration of readily available cellulose decreases and this activates the chitinase genes of Trichoderma sp. Which, in turn, produce chitinase to parasitize R. solani.

2. Antibiosis: Many microbes produce and secrete one or more compounds with antibiotic activity. In a general definition antibiotics are microbial toxins that can, at low concentrations, poison or kill other microorganisms. It has been shown that some antibiotics produced by microorganisms are particularly effective against plant pathogens and the diseases they cause. In all cases, the antibiotics have been shown to be particularly effective at suppressing growth of the target pathogen in vitro and/or in situ conditions. An effective antibiotic must be produced in sufficient quantities (dose) near the pathogen. In situ production of antibiotics by several different biocontrol agents has been studied. While several procedures have been developed to ascertain when and where biocontrol agents may produce antibiotics detecting expression in the infection court is difficult because of the heterogenous distribution of plant-associated microbes and the potential sites of infection.

However, in some cases, the relative importance of antibiotic production by biocontrol bacteria has been demonstrated. For example, mutant strains incapable of producing phenazines or phloroglucinols have been shown to be equally capable of colonizing the rhizosphere, but much less capable of suppressing soil borne root diseases than the corresponding wild-type and complemented mutant strains. Many biocontrol strains have been shown to produce multiple antibiotics which can suppress one or more pathogens. The ability of production of several antibiotics probably results in suppression of diverse microbial competitors and plant pathogens.

3. Metabolite production: Many biocontrol active microorganisms produce other metabolites that can interfere with pathogen growth and activities. Lytic enzymes are among these metabolites that can break down polymeric compounds, including chitin, proteins, cellulose, hemicellulose and DNA. Studies have shown that some of these metabolites can sometimes directly result in the suppression of plant pathogens. For example, control of Sclerotium rolfsii by Serratia marcescens appeared to be mediated by chitinase expression. It seems more likely that antagonistic activities of these metabolites are indicative of the need to degrade complex polymers in order to obtain carbon nutrition. Microorganisms that show a preference in colonizing and suppression of plant pathogens might be classified as biocontrol agents. For example, Lysobacter and Myxobacteria that produce lytic enzymes have been shown to be effective against some plant pathogenic fungi.

Studies have shown that some products of lytic enzyme activity may have indirect efficacy against plant pathogens. For example, oligosaccharides derived from fungal cell walls have been shown to induce plant host defenses. It is believed that the effectiveness of the above compounds against plant pathogens is dependent on the composition and carbon and nitrogen sources of the soil and rhizosphere. For example, in post-harvest disease control, addition of chitosan which is a non-toxic and biodegradable polymer of beta-1, 4-glucosamine produced from chitin by alkaline deacylation stimulated microbial degradation of pathogens. Amendment of plant growth substratum with chitosan suppressed the root rot caused by Fusarium oxysporum f. sp. radicis-lycopersici in tomato.

In addition to the above-mentioned metabolites, other microbial byproducts may also play important roles in plant disease biocontrol. For example, Hydrogen cyanide (HCN) effectively blocks the cytochrome oxidase pathway and is highly toxic to all aerobic microorganisms at picomolar concentrations. The production of HCN by certain fluorescent pseudomonads is believed to be effective against plant pathogens. Results of some research studies in this regard have shown that P. fluorescens CHAo, an antagonistic bacterium, produces antibiotics including siderophores and HCN, but suppression of black rot of tobacco caused by Thielaviopsis basicola appeared to be due primarily to HCN production. In another study reported that volatile compounds such as ammonia produced by Enterobacter cloacae were involved in the suppression of cotton seedling damping-off caused by Pythium ultimum.

4. Competition: The nutrient sources in the soil and rhizosphere are frequently not sufficient for microorganisms. For a successful colonization of phytosphere and rhizosphere a microbe must effectively compete for the available nutrients. On plant surfaces, host-supplied nutrients include exudates, leachates, or senesced tissue. In addition to these, nutrients can also be obtained from waste products of other organisms such as insects and the soil. This is a general believe that competition between pathogens and non-pathogens for nutrient resources is an important issue in biocontrol. It is also believed that competition for nutrients is more critical for soil borne pathogens, including Fusarium and Pythium species that infect through mycelial contact than foliar pathogens that germinate directly on plant surfaces and infect through appressoria and infection pegs. Results of a study by Anderson et al. revealed that production of a particular plant glycoprotein called agglutinin was correlated with potential of Pseudomonas putida to colonize the root system. P. putida mutants deficient in this ability exhibited reduced capacity to colonize the rhizosphere and a corresponding reduction in Fusarium wilt suppression in cucumber.

It has been shown that non-pathogenic plant-associated microorganisms generally protect the plant by rapid colonization and thereby exhausting the limited available substrates so that none are available for pathogens to grow. For example, effective catabolism of nutrients in the spermosphere has been identified as a mechanism contributing to the suppression of Pythium ultimum by Enterobacter cloacae. At the same time, these microbes produce metabolites that are effective in suppression of pathogens. These microbes colonize the sites where water and carbon-containing nutrients are most readily available, such as exit points of secondary roots, damaged epidermal cells and nectaries and utilize the root mucilage.

Competition for rare but essential micronutrients, such as iron, has also been shown to be important in biological disease control. Iron is extremely limited in the rhizosphere, depending on soil pH. In highly oxidized and aerated soil, iron is present in ferric form, which is insoluble in water and the concentration may be extremely low. This very low concentration can not support the growth of microorganisms. To survive in such environment, organisms were found to secrete iron-binding ligands called Siderophores having high ability to obtain iron from the micro-organisms. Almost all microorganisms produce siderophores, of either the catechol type or hydroxamate type.

A direct correlation was established in vitro between siderophore synthesis in fluorescent pseudomonads and their capacity to inhibit germination of chlamydospores of F. oxysporum. It was shown that mutants incapable of producing some siderophores, such as pyoverdine, were reduced in their capacity to suppress different plant pathogens. The increased efficiency in iron uptake of the commensal microorganisms is thought to be a critical factor in their root colonization ability which is a major factor in biocontrol performance of bacterial antagonists.

5. Induction of resistance: Plants actively respond to a variety of environmental stimulating factors, including gravity, light, temperature, physical stress, water and nutrient availability and chemicals produced by soil and plant associated microorganisms. Such stimuli can either induce or condition plant host defenses through biochemical changes that enhance resistance against subsequent infection by a variety of pathogens. Induction of host defenses can be local and/or systemic in nature, depending on the type, source and amount of stimulation agents.

Recently, plant pathologists have begun to characterize the determinants and pathways of induced resistance stimulated by biological control agents and other non-pathogenic microorganisms. The first pathway called Systemic Acquired Resistance (SAR), is mediated by Salicylic Acid (SA), a chemical compound which is usually produced after pathogen infection and typically leads to the expression of Pathogenesis-related (PR) proteins. These PR proteins include a variety of enzymes some of which may act directly to lyse invading cells, reinforce cell wall boundaries to resist infections, or induce localized cell death.

Second pathway, called Induced Systemic Resistance (ISR), is mediated by Jasmonic Acid (JA) and/or ethylene, which are produced following applications of some nonpathogenic rhizobacteria. Interestingly, the SA- and JA- dependent defense pathways can be mutually antagonistic and some bacterial pathogens take advantage of this to overcome the SAR. For example, pathogenic strains of Pseudomonas syringae produce coronatine, which is similar to JA, to overcome the SA-mediated pathway. Since the various host-resistance pathways can be activated to variable degrees by different microorganisms and insect feeding, it is therefore possible that multiple stimuli are constantly being received and processed by the plant. Thus, the magnitude and duration of host defense induction will likely vary over time. Only if induction can be controlled, i.e., by overwhelming or synergistically interacting with endogenous signals, will host resistance be increased.

Some strains of root-colonizing microorganisms have been identified as potential elicitors of plant host defenses. For example, some biocontrol active strains of Pseudomonas sp. and Trichoderma sp. are known to strongly induce plant host defenses. In other instances, inoculation with Plant Growth Promoting Rhizobacteria (PGPR) have been shown to be effective in controlling multiple diseases caused by different fungal pathogens, including anthracnose. A number of chemical elicitors of SAR and ISR such as salicylic acid, siderophore, lipopolysaccharides and 2, 3-butanediol may be produced by the PGPR strains upon inoculation.

A substantial number of microbial products have been reported to elicit host defenses, indicating that host defenses are likely stimulated continually during the plant's lifecycle. These inducers include lipopolysaccharides and flagellin from Gram-negative bacteria; cold shock proteins of diverse bacteria; transglutaminase, elicitins and a-glucans in Oomycetes; invertase in yeast; chitin and ergosterol in all fungi; and xylanase in Trichoderma. These findings indicate that plants would detect the composition of their plant-associated microbial communities and respond to changes in the quantity, quality and localization of many different signals. The importance of such interactions is indicated by the fact that further induction of host resistance pathways, by chemical and microbiological inducers, is not always effective in improving plant health or productivity in the field.

Methods of Application of Antagonists

1. Overall application: Successful application of biological control strategies requires more knowledge-intensive management. Understanding when and where biological control of plant pathogens can be profitable, requires an appreciation of its place within integrated pest management systems.

In general, the foundation of a sound pest and disease management program in an annual cropping system begins with cultural practices that alter the farm landscape to promote crop health. These include crop rotations that limit the availability of host material used by plant pathogens. Proper use of tillage can disrupt pathogen life cycles and prepare seed beds of optimal moisture and bulk density. Careful management of soil fertility and moisture can also limit plant diseases by minimizing plant stress. In nurseries and greenhouses environmental control can be more tightly regulated in terms of temperature, light, moisture and soil composition, but the design of such systems cannot wholly eliminate disease problems.

The second layer of defense against pests consists of the quality of crop germplasm. Breeding for pathogen resistance including fungal pathogens contributes substantially to crop success in most regions. Newer technologies that directly incorporate genes into crop genomes, commonly referred to as genetic modification or genetic engineering, are bringing new traits into crop. Other technologies, such as seed washing, testing for pathogens and treatments are also used to keep germplasm pathogen-free. In perennial cropping systems, such as orchards and forests, germplasm quality may be more important than cultural practices, because rotation and tillage cannot be used as regularly. Upon these two layers, growers can further reduce pathogen pressure by considering both biological and chemical inputs.

Biologically based inputs such as microbial fungicides can be used to interfere with pathogen activities. Registered biofungicides are generally labeled with short reentry intervals and pre-harvest intervals, giving greater flexibility to growers who need to balance their operational requirements and disease management goals. When living microorganisms are introduced, they may also augment natural beneficial populations to further reduce the damage caused by targeted pathogens.

2. Applying to the infection site: Application directly to the infection court at a high population level to swamp the pathogen (inundate application), seed coating and treatment with antagonistic fungi and bacteria, e.g. Trichoderma harzianum and Psudomonas fluorescens, antagonists applied to fruit for protection in storage, e.g. Pseudomonas fluorescens and application to soil at the site of seed placement. These types of applications are the most commonly used procedures which have resulted in the successful control of several fungal plant pathogens.

3. One place application: in this procedure, biocontrol microorganisms are applied at one place (each crop year), but at lower populations which then multiply and spread to other plant parts and give protection (augmentative application) against fungal pathogens. An Example of this method is Plant Growth Promoting Rhizobacteria (PGPR) and atoxigenic Aspergillus flavus on wheat seed scattered on the soil to spread to cotton flowers where they displace aflatoxin producing strains of A. flavus and fungal antagonists added to soil.

4. Occasional application: One time or occasional application maintains pathogen populations below threshold levels. In theory, parasites of the pathogen, or hypovirulent (disease carrying) strains of the pathogen, might be used and not require yearly repetition in which host plant is inoculated with attenuated strains of pathogenic that protects the host plant against the virulent strains of pathogen.

The use of Compost as Biofertilizer

Research data and observations in nurseries have shown that addition of composted organic matter to potting mixes results in suppression of soil borne fungal diseases. The concentration of suppressive microorganisms in compost amended substrates is very high, but greatly reduced in soils or potting mixes after the amendment. As a result, predictive disease suppression models have been developed based on the composition and concentration of microbial biomass.

The effectiveness of composts in suppression of soil borne diseases is dependent on heat kill, organic matters decomposition, recolonization of compost by suppressive microorganisms following heat kill and physical and chemical factors. Although previous works have focused on plant soil borne diseases, current research indicates that potting mixes containing composted organic materials which also have been inoculated with Trichoderma hamatum can be effective as a biocontrol alternative to foliar fungicides; however, the mechanism of this systemic type of induced resistance is not yet understood. Although the growers have traditionally relied on aged pine bark and composted biosolids to provide the potential for disease suppression, research indicates that composted animal manure have the potential to replace some of these components, but a consistent quantity and quality of these materials will need to be incorporated. The maturity (stability) of the composted manure and its salinity largely determine its ability to induce suppression.

Commercialization of Biocontrol

Commercial use and application of biological disease control have been slow mainly due to their variable performances under different environmental conditions in the field. Many biocontrol agents perform well in the laboratory and green house conditions but fail to do so in the field. This problem can only be solved by better understanding of the environmental parameters that affect biocontrol agents. In addition to this problem, there has also been relatively little investment in the development and production of commercial formulation of biocontrol-active microorganisms probably due to the cost of developing, testing, registering and marketing of these products.

Biological control agents are generally formulated as wetable powders, dusts, granules and aqueous or oil-based liquid products using different mineral and organic carriers.

Currently in the market, a number of biologically based products are being sold for the control of fungal plant diseases. A growing number of companies are also developing new products that are in the process of registration. Many of these companies are small, privately owned firms with a limited product-line. Others are publicly traded and have substantial capitalization values. In addition, larger companies with more diverse product lines that include a variety of agrochemicals and biotechnological products have played a significant role in the development and marketing of products for the control of plant pathogens.

Biocontrol products are either marketed as stand-alone products or formulated as mixtures with other microbials. Some products with biocontrol properties may not be registered, but are sold instead as plant strengtheners or growth promoters without any specific claims regarding disease control. To help improve the global market perception of biopesticides as effective products, the biopesticide Industry Alliance is establishing a certification process to ensure industry standards for efficacy, quality and consistency. To improve commercial use and application of biological disease control it is extremely important to emphasize and concentrate on several factors including training of growers, formulation of biocontrol microorganisms and studying the role of environmental factors.

Conservation Biological Control

Conservation biological control is one of biological control main branches, which can be first realized by reducing the use of pesticides, use of selective pesticides, careful timing and placement of pesticide applications. We have seen what happens when insecticides destroy the natural enemies of potential pests. Insects that were of little economic importance may become destructive pests. When nontoxic control method is used natural enemies are more likely to survive and reduce the populations of pests.

During non-crop periods, natural enemies may need of benefit from pollen, nectar or honeydew (produced by aphids). Many crop plants flower for only short time, so flowering plants along the edges of the field or within the field may be needed for pollen and nectar. Preservation of natural enemies can be achieved by providing habitat and resources for natural enemies. They are generally not active during the winter. Unless they are re-released each year, they must have a suitable environment for overwintering. They usually pass the winter in crop residues, other vegetation or in the soil. Ground cover of fruit orchards, winter crops (like alfa alfa and breccias), usually provides shelter for overwintering natural enemies. Adding plants or other food sources for natural enemies must be done with knowledge of the behaviour and biology of the natural enemy and the pest.

It is widely known that the simplifications of agriculture systems towards monoculturing are mainly responsible for decreasing environmental quality, threatening biodiversity and increasing the possibility of insect outbreaks. Modern crops are often monocultures in highly specialized production units, where not only crop cultivation but also harvest and packaging techniques are specialized. The development of farming systems (field or landscape) with greater dependence on ecosystem services, such as biological control of insect pests, should increase the sustainability of agro-ecosystems. Farming systems like greenhouses, annual crop systems and other practices that end with removing the whole crop after harvesting, may give rise to elimination of biodiversity, and decreasing the population of natural enemies in the fields or in different agricultural environments, as appeared in figure below. Collection and transferring of natural enemies to environmentally controlled habitats could be useful in utilizing these natural enemies until releasing them in the next crop season.

Complete removal of maize may eliminate natural enemies (A) or after roses cutting (B).

Thus, they will try to contribute to preserve the natural biodiversity in the agricultural environment and provide natural alternatives to chemical pesticides. We concentrate here on the effects of conservative biological control on NE biodiversity and cleanliness of environment.

- Collection of natural enemies before the end of crop season.

- Preservation of collected natural enemies in special greenhouses during non-crop periods.

- Releasing the preserved natural enemies on target crops in the next growing season.

- The sequence of these practices is illustrated in figure below.

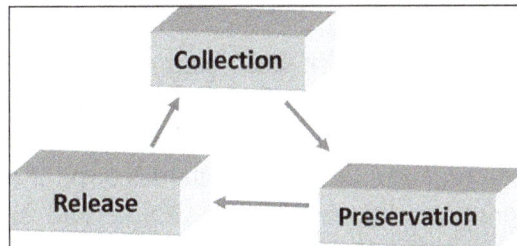

Logical practices diagram of conservation biological control.

Collection of Natural Enemies

The first step of the suggested strategy is collection of NE from fields shortly before the complete removal of plants and disappearance of occurring NEs. At the end of the crop season, the NEs are usually in their top population densities.

Collection Time

Summer collection: High numbers of natural enemies may be found during the growing season on areas cultivated with some crops. These crops may not be in need for these natural enemies especially in absence of insect hosts or preys. For example, after heavy infestation of aphids to maize plants, high populations of aphid predators (lacewings and lady beetles) are built up. These predators could be mass collected and directly transferred to the preservation greenhouses or directly to other target crops that are in need for them.

Autumn collection: Before the end of most of annual crops, there are huge numbers of natural enemies which may be lost after harvesting and removing the plants. These NEs could be collected, preserved in greenhouses during non-crop periods then released in the next season.

Winter collection: In cases of permanent crops like fruit orchards and alfa alfa during cold weather in winter, many numbers of natural enemies may be lost as a result of absence of their hosts and preys, especially during non-suitable weather conditions. These natural enemies could be collected and transferred to greenhouses where maintained and improved them in numbers and quality control until release during the next crop season.

Collection sites

Natural enemies may be abundant in many sites around the year including landscape, fruit orchards, vegetable and field crops and ornamentals and others.

Collection Techniques

Collection techniques differ according to the nature of natural enemies, crop, time and site.

The common collection techniques are vacuum collection, sweeping net, pitfall traps, manual collection etc.

Collection techniques depend on many factors like pest species, host plant, type of natural enemy, habit, time, weather and others.

Picking up Infested Plant Leaves

Plant leaves are picked up and transferred in cloth bags to the preservation greenhouses where emerged natural enemies could be classified and maintained. Infested leaves containing parasitized insects of mulberry trees were picked up and transferred to the laboratory; then the parasitoids were counted after their emergence. Leaves of tomatoes or potatoes infested with leaf feeders Phthorimaea operculella, Spodoptera littoralis, Tuta absoluta and Agrotis ipsilon were collected then the parasitoids were counted after their emergence. Immature predators were collected and transferred to the laboratory together with the plant material infested by their prey scale insects for rearing to the adult stage.

Beating Tree Branches in Cloth Bags

Leaves and/or branches (shoots) are picked up from trees and beaten in cloth or paper bags; then they were transferred to preservation greenhouses. Hendawy et al. used this method for sampling predators and parasitoids of mealybug on mulberry trees. Small branches of pine trees were beaten in cloth bags and transferred in the laboratory for surveying mealybug natural enemies. Also mango trees were sampled by the same methods for monitoring the natural enemies of Aulacaspis tubercularis and Kilifia acuminata. Infested small branches were collected in cloth bags and predators were counted in the laboratory.

Sweeping Net Technique

Sweeping net technique is a common technique for collecting parasitoids and predators such as Chrysopid, Syrphid and Coccinellid species from vegetable and field crop plants. Sayed, ELbehery, and Badr used the sweeping net in tomato or potato fields, usually by 50 double strikes by walking diagonally across the experimental plots.

Direct Collection of Insect Individuals

Parasitized caterpillars or white grubs infesting roots are directly collected and transferred to preservation greenhouses where emerged parasitoids could be classified and maintained until their releases in the next season. Sallam et al. collected white grubs infesting sugarcane roots and reared until parasitoid emergence. Larvae of armyworms were collected in sugarcane fields and were taken to the laboratory and fed on pieces of cane leaves until parasitoid emergence.

Aspirator Devices

Aspirator or vacuum devices are used for collecting flying natural enemies from trees, orchards, vegetable and field crops. Adult parasitoids and predators were collected using an aspirator and dropped into a jar. Erler and Tunç used aspirator devices for collecting the predacious mites from orchards and wild trees.

Preservation of Natural Enemies

Preservation greenhouses are dedicated for natural enemies rather than commercial production of crops. Preservation practices represent the cornerstone of conservation biological control. Preservation practices could be applied individually or in combination to maintain and improve efficiency of collected natural enemies.

Practices of preservation of natural enemies are many and vary according the types of natural enemies, the target pests, the plants and the ecological conditions.

Plant-provided Food

Many plants can provide food sources for natural enemies like nectar, pollen and plant sap but the effect of these food sources depends on the type of predator/parasitoid. Specialist natural enemies reproduce only in the presence of their specific prey/host species. However, most other natural enemies are feeding on both plant resources and prey. Wäckers et al. stated that adults of parasitoids and gall midges can increase their longevity, flight activity and oviposition by feeding on nectar. General predators consume multiple prey types and may feed also on nectar and pollen provided by plants. Adding some flowering plants like sweet alyssum and coriander to a sweet pepper crop resulted in higher densities of hoverflies. Plants that produce a lot of pollen, like Ricinus communis, provided more pollen to predatory mites. Flowering alyssum provided food resources for the predatory bugs Orius laevigatus and Orius majuscules during times of prey scarcity. Flowering ornamental pepper can support and increase populations of Orius insidiosus in ornamental crops. Another approach can be to select crop varieties with increased levels of plant-provide food resources. Thus, the availability of plant-provided food can be a driving force in biocontrol success program.

Food Sprays

Artificial or natural food supplements can be sprayed or dusted onto the crop to support natural enemies in crops where nectar and pollen are absent or only present at low densities. For example, pollen sprays can serve as food for predatory mites and enhance their efficacy against thrips and whiteflies on cucumber. Corn pollen is also suitable for increasing populations of Amblyseius swirskii and Euseius scutalis. These pollens could be mechanically collected in large quantities. Other types of pollen are commercially available for pollination, such as apple pollen and date palm pollen. Application of pollen on chrysanthemum plants increases the establishment of many natural enemies. Studies with predatory mites showed that adding Typha latifolia pollen to a crop clearly enhanced the biological control of thrips, even though the pollen is edible for thrips itself. The development of inexpensive alternative food sources is one of the major opportunities and challenges for enhancing biological control in different crop.

Introducing Non-crop Plants Harbouring the Prey Species

The use of alternative prey/host plant species for the preservation of released natural enemies in many crops has been of interest for biological control of insect pests. A widely applied system in different crops has been the use of monocotyledonous plants with cereal aphids that serve as alternative hosts for parasitoids of aphids that attack the dicotyledon crop. Prey/host plants can also be established on the edges of the field to bridge non-crop periods and contribute to the preservation of natural enemies. Some alternative prey species that are not harmful to the crop may support their natural enemies. Woody habitats (hedgerows, field margins) often provide a more moderate microclimate than the centre of fields, protecting natural enemies against extreme temperature variations.

Applying Artificial Food for Natural Enemies

The application of yeast and sugars in chrysanthemum maintained populations of astigmatic mites that are suitable prey for phytoseiid predatory mites.

Artificial Field Rearing Units

Rearing natural enemies in controlled conditions has been developed into artificial rearing units for some natural enemies. For example, rearing sachets containing bran with saprophytic fungi for feeding astigmatic mites (prey) were used for rearing predatory mites. Many modifications with different types of preys, predatory mites, food sources for astigmatic mites such as sugars, starch, yeast and types of sachets have been developed. Such units may produce predatory mites for 3–6 weeks. This could be optimized by balancing the rate of predator, prey and food in the rearing unit.

Inoculation with Low Pest Levels

A risky method to support natural enemies is the release low levels of pest species into crops. Inoculating plants with a low level of spider mites early in the growing season and release predators afterwards enhanced the establishment of predatory mites in the crop. Currently, this method is mainly used in sweet pepper crops. Thus, allowing low levels of pests, in numbers insufficient to cause crop damage, might contribute to natural enemies preservation.

Supplementing Mixed Diet for Natural Enemies

The population of natural enemies in crops can be increased by providing mixed diets of prey and/ or non-prey food sources. Survival and reproduction of O. insidiosus were enhanced when aphids with thrips were supplemented as a prey source. Supplementing thrips with pollen increased egg production of O. laevigatus and predation rates of thrips larvae. Thus, supplementing diets of single pest species for predators with alternative prey or food may increase predator population and enhance biological control.

Providing Oviposition Sites and Shelters

Suitable oviposition sites are essential for reproduction of many predators. Orius spp. and Mimulus pygmaeus lay their eggs into soft plant parts and ovipositional acceptance of the host plant depends on the morphological characteristics such as epidermal thickness or trichome density. The hard plant parts are not very suitable for oviposition behaviour of predators and may disrupt their establishment. Cutting soft stems of flowers may remove a potential new generation of natural enemies from the fields. The same problem can also occur on tomato with the de-leafing practice that has a strong negative effect on the development of mired predator populations and Encarsia formosa by removing parasitized whitefly scales. These problems may be solved by adapting the de-leafing strategy or providing host plants with suitable oviposition sites for natural enemies.

A number of plants are considered as refuges for natural enemies. For example, the vein axils of sweet pepper plants are used by predatory mites for oviposition which reduced cannibalism and increased survival by providing such suitable microclimate. Adding Viburnum tinus and Vitis riparia plants in roses enhanced mite control by predatory mites.

Planting Suitable Non-crop Plants near Fields

Mirid predators often migrate from non-crop plants into tomato fields, where they add to the control whiteflies, leaf miners and T. absoluta . The natural existence of predatory bugs in tomato fields seems to be strongly related to the surrounding landscape. Migration of Orius spp. from neighbouring wild plants into sweet pepper fields may compete with populations of released O. laevigatus. Many studies suggested that preservation biological control of predators can be enhanced by planting suitable non-crop plants near fields either to support migration into the crop or to provide a shelter when field crops are harvested and plants removed. Field surroundings may also contribute to the migration of parasitoids into fields. Providing overwintering shelters may enhance lacewings by providing diapausing adults with artificial overwintering chambers in greenhouses. These methods may contribute to early establishment of natural enemies in new season in the spring.

Induced Plant Responses

Induced plant resistance against insects includes direct traits, such as the production of toxins and feeding deterrents that reduce survival, host preference, fecundity or developmental rate of pests and indirect traits, which attract and/or retain natural enemies. The latter contains traits such as the plant producing volatiles and floral nectar. Insect-induced plant volatiles help natural enemies to detect their prey/hosts in a crop, whereas floral nectar production is increased in response to insect attack, guiding natural enemies to find their prey/hosts. Preservation of natural enemies might be enhanced in different crops by breeding varieties that produce more volatiles and nectar.

Applying Semiochemicals

Behaviour of natural enemies is directed by semiochemicals. Attraction of natural enemies with synthetic compounds, similar to plant volatiles, is being tested in crops. Natural enemies may also respond to odours that are produced by their prey/host species, such as sex pheromones or alarm pheromones. Sex pheromones are used either to monitor or mass trapping pest populations. However, volatiles for improving natural enemy performance are so far not applied in many crops. Glinwood et al. mentioned that pheromones could be used to treat clusters of aphid infested plants in fields, which might increase efficacy of released parasitoids. Lures may also be used to attract released natural enemies in order to help them establish. Applying attractants in combination with food sprays may promote oviposition of released chrysopid predators into the target crop. Hexane extract of corn borer larvae was applied on corn plants to enhance performance of larval parasitoid Bracon brevicornis adults against the corn borers Ostrinia nubilalis and Sesamia cretica.

Pesticide Side-effects

Preservation of natural enemies should not be combined with pesticides, as most pesticides have lethal effects on NEs. Mitigation of side-effects on preservation of natural enemies can be realized by selecting pesticides that are compatible as possible with natural enemies.

Finally, with transfer of collected natural enemies into greenhouse with environmentally safe conditions, where these natural enemies can be fed on the pollen and nectar of flowering crops (clover and alfa alfa), these plants will provide shelter for the natural enemies. This procedure will be continued until the next crop season, where the proper site and time of release.

Balzan and Moonen mentioned that studying field margin vegetation enhances biological control agents in addition to crop damage suppression from many insect pests in tomato fields. They suggested that these habitats may be important during early crop colonization by natural enemies. These results indicate that the inclusion of flower strips enhances the preservation of arthropod functional diversity in ephemeral crops, and that diverse mechanisms are important for controlling different pests. However, the efficiency of habitat management is likely to be better when it is complemented with the preservation of diverse seminatural vegetation in the pre-existing field margin. Therefore, the field margin should be considered and evaluated before the inundative release strategy.

Release of Natural Enemies

Release techniques are varied according the type of biocontrol agents, host plants, weather conditions. For example, egg parasitoids are released as parasitized egg patches; larval parasitoids are released as adults. Predators are usually released in the pupal stage. Timing, rate and frequency of release are determined according to the nature of the target pests, natural enemies and crops. Pathogens like entompathogenic nematodes could be applied as sprays or injection.

Egg Parasitoids

The common techniques of releasing egg parasitoids are paper cards or strips holding the parasitized eggs. Cardboard strips containing parasitized eggs in tubes were released in tomatoes for controlling T. absoluta. Trichogramma buesi was released against Pieris rapae eggs in cabbage fields. A dose of 3000 Trichogramma evanescens wasps/card x three cards/tree was applied; each card contains three different ages of Trichogramma to keep searching adults continuously; 8–11 releases were performed per year at 2-week intervals against Prays oleae in olive fields . Five releases of Trichogramma at two release levels (50 and 75 cards/ha, each contains 1000 parasitoids) were released in grape orchards for controlling Lobesia botrana. Over 100,000 parasitoids per Feddan were released against Chilo agamemnon in sugarcane fields; five releases were applied during season.

Bollworms are causing highly infested boll in cotton; Trichogramma were applied for control them. Different releasing Trichogramma in cards, each contain 1000 parasitoid for several times. Four Trichogramma species (T. japonicum, T. chilonis, T. dendrolimi and T. ostriniae) was evaluated against Chilo suppressalis in rice fields. T. chilonis parasitized more eggs, while T. dendrolimi and T. japonicum performed the best.

Larval Parasitoids

Larval and pupal parasitoids are released in the pupal stage. Parasitized pupae just before emergence are carried on special carriers like talc powder and distributed in the target fields. Releasing Bracon spp to control corn borer larvae is one of the effective methods for controlling such insects. Two ectoparasitoid species Bracon sp. and Necremnus sp. were released in tomatoes. Necremnus sp. Nrartynes and other braconid species have already been proved to be potential key biocontrol agents of T. absoluta in tomato field.

White Fly Parasitoids

Encarsia spp. or Eretmocerus spp. are released as parasitized pupae shortly before adult emergence. Additional Encarsia species have been released against Bemisia tabaci; reached to 65% parasitized whiteflies . Simmons and Abd-Rabou confirmed that inundative releases of parasitoid Eretmocerus mundus against B. tabaci into tomato and cotton fields increased parasitization rates. Findings from their research may be useful in the enhancement and preservation of parasitoids of Bemisia.

Aphid Parasitoids

Aphid parasitoids are released as parasitized mummies of aphid host. Semi-field experiments were carried out to evaluate the performance of releasing parasitoid species Diaeretiella rapae for controlling Brevicoryne brassicae, Aphis craccivora and Aphis nerii infesting cabbage, faba bean and oleander plants. The highest percentage of parasitism was 92.20, 83.20 and 79.30% for D. rapae at 20 parasitoids/200aphids per cage in semi-field test B. brassicae, A. craccivora and A. nerii, respectively. The maximum numbers of mummies in the field were 185.60, 166.4 and 158.6 for D. rapae at 20 parasitoids per cage and minimum of 124.60, 97.40 and 83.0 mummies at five adults per cage.

Parasitoids of Scale Insects

Parasitoids of scale insects are released as parasitized host individuals. About 953,000 of Coccophagus scutellaris as parasitized individuals were released and evaluated for controlling soft scale insects Ceroplastes rusci on citrus, Ceroplastes floridensis on citrus, Coccus hesperidum on guava, Pulvinaria floccifera on mango, Pulvinaria psidii on mango, Saissetia coffeae on olive and Saissetia oleae on olive. The population of parasitoid C. scutellaris showed a significant correlation with the build-up of the population of the soft scale insects population in all of the release orchards studied.

Mealybug Parasitoids

Parasitoids of mealybug are released as parasitized host individuals. Anagyrus kamali and Gyranusoidea indica were released at ten sites on ornamental plants. 300,000 parasitoids of A. kamali were released to control Maconellicoccus hirsutus. Population density of M. hirsutus was reduced by approximately 95% and A. kamali was the predominant parasitoid.

Predators of T. Absoluta and B. Tabaci

General predators (lacewings and lady beetles) are released in the pupal stage with the suitable carriers. These general predators are used commercially for regulating many insect and mite pests. Nesidiocoris tenuis and M. pygmaeus were also released and caused a significantly reducing T. absoluta and B. tabaci populations.

Predacious Mites

Individuals of predacious mites carried on special materials are released for regulating spider mites and whiteflies in tomato and pepper in the greenhouses.

Combination Entomopathogenic Nematodes (EPNs) and Egg Parasitoid

Natural enemies may be released in integration with each other to regulate one or set of insect pests. Entomopathogenic nematodes (EPNs) and Trichogramma were used for S. cretica, C. agamemnon and O. nubilalis, respectively, in corn fields. The infested plants S. cretica were sprayed one time with 500 and 1000 IJs/ml of Steinernema carpocapsae and Heterorhabditis bacteriophora. Three releases of T. evanescens were conducted to control C. agamemnon and O. nubilalis.

Entomopathogenic Nematods Application

Entomopathogenic nematods are injected in tunnels made by the red palm weevil larvae or sprayed around the trunks of infested trees to control the pest adults.

Augmentation or Inundative Biological Control

"Biological control" means the purchase and release of beneficial natural enemies to control insect and mite pests. This approach is known as augmentation of natural enemies. The underlying reason for the widespread recognition of this technique is that it relies on the use of commercial products that are advertised in farming and gardening magazines and publicized in the media. Further, the historical use of pesticides has trained us to think about pest management in the context of purchased products. However, of the three general approaches to insect biological control, augmentation is the least sustainable because it does require the regular or periodic investment in purchased inputs. Nonetheless, in some pest situations it is a highly efficacious, cost effective, and environmentally sound approach to pest management.

The practice of augmentation is based on the knowledge or assumption that in some situations there are not adequate numbers or species of natural enemies to provide optimal biological control, but that the numbers can be increased (and control improved) by releases. This requires a readily available source of large numbers of natural enemies which has fostered the development of companies to produce and sell these. Many companies (called insectaries) produce a variety of predatory and parasitic insects; other companies produce and market insect pathogens for use as microbial insecticides.

There are two general approaches to augmentation:

1. Inundative releases and inoculative releases: Inundation involves releasing large numbers of natural enemies for immediate reduction of a damaging or near-damaging pest population. It is a corrective measure; the expected outcome is immediate pest control. Because of the nature of natural enemy activity, and the cost of purchasing them, this approach using predaceous and parasitic insects is recommended only in certain situations, such as the mass release of the egg parasite Trichogramma for controlling the eggs of various types of moths. The utilization of some microbial insecticides (such as those containing Bacillus thuringiensis) is also inundation. Inoculation involves releasing small numbers of natural enemies at prescribed intervals throughout the pest period, starting when the pest population is very low. The natural enemies are expected to reproduce themselves to provide more long-term control. The expected outcome of inoculative releases is to keep the pest at low numbers, never allowing it to approach an economic injury level; therefore, it is more of a preventive measure. Two examples are the release of predatory mites to protect greenhouse crops, and the inoculation of soils with the milkyspore pathogen (Bacillus popillae) to control Japanese beetle grubs.

2. Targets of augmentation: Augmentative biological controls have not been developed for all pest problems. Indeed, relatively few situations are amenable to this approach. One of the most frequent uses of augmentation is to protect greenhouse crops, a practice that was started in Europe over 30 years ago in response to widespread occurrence of insecticide resistance in greenhouse pests. Today, commercial natural enemies are available for controlling aphids, mites, scale insects, mealybugs, leafminers, thrips, caterpillars, and other greenhouse pests.

Another situation that uses augmentation is the control of filth flies in livestock manure. Several parasites are commercially available; their impact is heightened when used in conjunction with appropriate manure handling practices.

Augmentation, other than the use of microbial insecticides, has not been widely used in Midwest orchards and vineyards. It is heavily used in some areas of California, where cooperative, non-profit citrus protection districts have their own insectaries for natural enemy production. In row crops, generalist natural enemies are frequently used, such as the egg parasite Trichogramma, green lacewings, and microbial insecticides. In the United States, augmentation has probably been used the least on field crops, partly because of the lack of a complex of effective natural enemies, and partly because the expenses may not be acceptable on low-value crops. Bacillus thuringiensis is commonly used for controlling European corn borer, and considerable research is aimed at making the releases of Trichogramma, also for corn borer, a viable option. Home gardeners are increasingly using natural enemies to protect food crops and landscape plants. There are several other areas where commercial natural enemies may be used, and some companies target specialized markets, such as gypsy moth, fire ant, and stored product pests.

3. Types of natural enemies available: There are over 100 types of commercially available natural enemies, including predatory insects and mites, parasitic insects, insect-parasitic nematodes, and insect pathogens. Although this sounds like a high number, it is small compared to the total number of pests in the United States. Further, many of these natural enemies are specialized for pests on crops such as cotton and citrus that are not grown in the upper Midwest. Other commercial natural enemies, such as lady beetles and praying mantids, are of questionable value even though they have been highly popularized.

4. Efficacy: There is no doubt that well-researched applications of natural enemies can be very effective. This includes the use of microbial insecticides as well as many specific uses of predators and parasitic insects. There is also no doubt that many natural enemies that are sold do not control the intended target pest(s). The reasons for the latter scenario are multiple and complex. They range from the ridiculous (my favorite example involves a community that purchased and released lady beetles for mosquito control) to the obscure. Probably the common thread that exists with "failures" is a lack knowledge. This encompasses both a lack of research needed to make recommendations for successful implementation, and a lack of needed knowledge on the part of the pest manager about the biology of the pests, the natural enemies, and their environment, all of which is crucial to making augmentation work. In this short space, our best advice for pest managers interested in embarking on a new augmentation program is to first get as much information as possible to assure a reasonable chance for success.

5. Cost effectiveness: Some natural enemies are much easier and less expensive to produce than others; this is reflected in their prices. Because of the differences in prices and usage patterns, it is hard to generalize on the cost effectiveness of purchased natural enemies. Other less obvious factors also have to be considered, especially when comparing the release of natural enemies to the use of pesticides. These include pesticide resistance management, worker protection, impacts on non-target pests, environmental considerations, and marketing practices (such as conventional vs. organic). Another problem is that, for many commercial natural enemies and their potential target pests, there is not adequate research to recommend specific release rates based upon pest population levels. There are, however, many situations where augmentative biological control is cost competitive with the use of pesticides or other pest management practices. The high value of many specialty crops reflects high production costs, including pest management. In such crops, the expense of biological control may be relatively low when compared to overall production costs. On low value crops, the use of natural enemies must be inexpensive to be justified. This does not preclude the use of augmentation in field crops; inundative controls such as Bacillus thuringiensis and Trichogramma may be cost effective, as can be inoculative releases that rely on relatively low numbers of natural enemies. The cost of natural enemy releases should be carefully evaluated, as with any other production cost.

Biological Control of Plant Diseases

Plant diseases are a major constraint on crop production in all agricultural and horticultural systems. All crops are susceptible to diseases caused by a variety of pathogens (bacteria, fungi, and viruses). In general losses of crops due to disease amount to 25% of world crop production per annum Lugtenberg. Of course, losses are not distributed evenly but in some cases may be much higher resulting in loss of the entire crop. At the very least this can have severe financial implications at the local, regional or national levels. Atworst it can lead to famine with considerable loss of life.

Management of plant diseases is a significant cost component in crop production. Traditionally the approaches to dealing with disease in agricultural ecosystems includes breeding resistant varieties of the crops species, hygiene to prevent the spread of contaminated soil or seed, and fungicides to kill potentially infecting fungi. However increasing concerns about the effects of fungicides in the environment and residues in food have resulted in deregistration of a number of fungicides. The need to replace these has increased interest in biological control of plant diseases in recent years. Biological control is the suppression of disease by the application of a Biocontrol Agent (BCA) usually a fungus, bacterium, or virus, or a mixture of these to the plant or the soil. The BCA acts to prevent infection of the plant by the pathogen, or establishment of the pathogen in the plant. The main advantage of using a BCA is that they are highly specific for a pathogen and hence are considered harmless to non-target species. Over the past decade there have been many reports of the identification of effective BCAs for fungal and bacterial diseases in crops and a number of BCAs are in commercial production. In recent years our understanding of how BCAs protect the plant from infection has changed dramatically with the application of genomics. In order to implement an effective biocontrol programit's essential to understand how BCAs work to prevent disease development.

Effectiveness of Biocontrol

The level of disease control achieved by application of BCAs to a crop can be close to or equivalent to that achieved by application of a fungicide. Application of a fungicide to Phytophthora cactorum infected apple resulted in 100% disease suppression whilst application of various BCAs singly resulted in levels of disease suppression between 79%–98% depending on the BCA. In another study application of a Bacillus amyloliquefaciens BCA to mandarin fruit suppressed P. digitatum infection by 77% which compares to 96% after application of the fungicide imazalil. The efficacy of a BCA can be enhanced by mixing with a fungicide provided the fungicide does not adversely affect the BCA. Infection of strawberry by Botrytis cinerea was reduced to low levels by application of a Trichoderma atroviridae BCA, but was eliminated by application of the BCA with a fungicide. Interestingly in this case the fungicide alone was less effective than the BCA alone. Nakayama and Sayama reported a similar enhancement in disease control using a BCA fungicide mix to inhibit powdery scab of potato. Where there are comparative results for disease suppression in glasshouse and field trials, the degree of suppression tends to be lower in the field trials e.g. in the study of Fu et al the degree of suppression was 24% lower in the field. This is considered to reflect the more diverse environment in the field. A number of studies have demonstrated that biocontrol can also be used effectively against postharvest diseases.

Some endophytes protect against multiple pathogens. An endophytic strain G3 with potential as a biocontrol agent was isolated from the stems of Triticum aestivum L. It was classified by 16S rDNA sequencing as a member of Serratia. Although strain G3 displayed a broad spectrum of antifungal activity in vitro against a number of phytopathogens such as Botrytis cinerea, Cryphonectria parasitica, Rhizoctonia cerealis and Valsa sordida is has not been tested for disease suppression. A strain of Bacillus pumilis isolated from the endosphere of poplar suppressed the growth of three pathogens Cytospora chrysosperma, Phomopsis macrospora and Fusicoccum aesculi in greenhouse tests.

Host Genotype Effect

One of the problems with biocontrol is the lack of consistency in disease suppression by a BCA. Differences in host genotype contribute to differences in responses to a BCA. In control of Phytophthora meadii infection of Hevea brasiliensis by Alcaligenes sp. the degree of control differed between two cultivars of the host. A cultivar effect was also observed in studies on biocontrol of diseases in strawberry and pepper. The specificity effect may be related to the production of plant molecules that activate transcriptional activators of the LuxR family in the bacterium. The products of the LuxR genes act as global regulators controlling such processes as biofilm formation and antibiotic production among others. Although LuxR regulators normally operate in quorum sensing systems whereby bacteria communicate with each other, some such as the PsoR gene of P. fluorescens and the OryR gene of Xanthomonas oryzae respond to plant compounds thereby facilitating plant-BCA communication. Alternatively communication could be mediated by secondary metabolites produced by the BCA. Endophytes produce a large array of different types of secondary metabolites many of which have not been detected directly but have been inferred from genomic analysis. There are examples where the synthesis of secondary metabolites stimulates changes in plant metabolite production and vice-versa.

Mixtures of BCAs

Several researchers have reported that using mixtures of BCAs has increased the consistency of biocontrol across sites with different conditions. In studies on infection of potato by Phytophthora capsici greater disease control was achieved using a mixture of three bacterial BCAs compared to using the single strains. Slininger et al. in their investigation into postharvest dry rot of potato found that formulations of mixed BCAs performed more consistently across 32 storage environments varying in cultivar, washing procedure, temperature, harvest year, and storage time. Enhanced biocontrol using mixtures of BCAs has been reported for control of late blight in potato, diseases of poplar, chilli, and cucumber. It is also possible that different mixtures may need to be used in different climatic areas. Thus there is a need to identify a number of potential biocontrol agents. Mixtures do not always give increased control. In some cases there may be antagonism between the BCAs that results in reduced control compared to single strains. In evaluating agents for control of fire blight in pear, Stockwell et al. found that mixtures of Pseudomonas fluorescens A506, Pantoea vagus C9–1 and Pantoea agglomerans Eh252 were less effective that the individual strains. It was found that the Pantoea strains exert their effect through the production of peptide antibiotics. In the mixture these were degraded by an extracellular protease produced by P. fluorescens A506. Roberts et al. have also reported antagonism between BCA strains. They observed that populations of Trichoderma virens GL3 or GL321 were both substantially reduced after co-incubation with Bacillus cepacia BC-1 or Serratia marcescens isolates N1–14 or N2–4 in cucumber rhizospheres. These reports highlight the importance of considering possible antagonism between strains when developing a biocontrol formulation. Co-cultivation in vitro can sometimes reveal inhibitory effects but not always. In the study by Stockwell et al. antagonism between the species in the mixture would not have been evident from co-cultivation of the three species, it would only have been evident if the mixture was tested in a confrontation assay with the pathogen.

Origin of BCAs

Most commonly BCAs are isolated by screening organisms from the rhizosphere or endophyte population for inhibition of growth of the target pathogen in vitro. Those that show inhibition are assessed further although it should be stressed that in vitro inhibition is not always a successful indicator of a successful BCA as there are other mechanisms of disease suppression (stimulation of host growth induction of host defence; occlusion of pathogen; competition for nutrients; toxin inactivation) that do not involve growth inhibition, and there are other characteristics required for a successful BCA such ability for mass production and persistence under field conditions. Prominent among those species of rhizosphere and endophytic bacteria that are effective BCAs are the actinomycetes and species from the genera Pseudomonas and Bacillus. Among the fungi that constitute effective BCAs species of the genus Trichoderma are well represented. All of these are capable of synthesizing an array of secondary metabolites.

Actinomycetes make very good BCAs. Endophytic actinobacteria isolated from healthy cereal plants were assessed for their ability to control fungal root pathogens of cereal crops both in vitro and in planta. Thirty eight strains belonging to the genera Streptomyces, Microbispora, Micromonospora, and Nocardioidies were assayed for their ability to produce antifungal compounds in vitro against Gaeumannomyces graminis var. tritici (Ggt), the causal agent of take-all disease in wheat, Rhizoctonia solani and Pythium spp. Spores of these strains were applied as coatings to wheat seed, with five replicates (25 plants), and assayed for the control of take-all disease in plantain steamed soil. The biocontrol activity of the 17 most active actinobacterial strains was tested further in a field soil naturally infested with take-all and Rhizoctonia. Sixty-four percent of this group of microorganisms exhibited antifungal activity in vitro, which is not unexpected as actinobacteria are recognized as prolific producers of bioactive secondary metabolites. Seventeen of the actinobacteria displayed statistically significant activity in planta against Ggt in the steamed soil bioassay. The active endophytes included a number of Streptomyces, as well as Microbispora and Nocardioides spp. and were also able to control the development of disease symptoms in treated plants exposed to Ggt and Rhizoctonia in the field soil.

Hypovirulent isolates of a pathogen species can also act as BCAs. A naturally occurring hypovirulent isolate of Phytophthora nicotianae was found to effectively control citrus root rot caused by P. nicotianae and P. palmivora. In another study binucleate isolates of Rhizoctonia solani were effective at controlling damping off diseases in pepper caused by Rhizoctonia solani or Pthyium ultimum. Hypovirulent isolates of the Chestnut Blight disease pathogen Cryphonectria parasitica were widely and successfully used to control the disease in chestnut trees.

Viruses as BCAs

Due to the paucity of effective bactericidal compounds for management of bacterial phytopathogens there is renewed interest in the use of bacterial viruses (bacteriophage or phage) as BCAs for control of bacterial diseases. Recent studies have demonstrated significant reduction in disease severity for a range of pathogens including, Agrobacterium, Xanthomonas, Ralstonia solanacearum, Erwinia amylovora, and Streptomyces on a variety of crops. The advantages of using phage are: a) ease of production; b) high specificity for the target organism; c) long shelf life. The phage can be grown in the field using an avirulent form of the pathogen infected with the phage applied as a dressing to the crop. The avirulent strain acts as a vehicle for production of the phage but is not able to damage the crop. In effect this creates a self-perpetuating biocontrol system in the field. In the studies on the suppression of Ralstonia wilt of tomato using phage, infective phage particles were detected four months after treatment. One problem associated with the use of phage BCAs is the development of resistance in the host bacterial population. The use of a cocktail containing a number of host range mutants is recommended to overcome this. Such mutants can be evolved in the lab. The persistence of phage BCAs in the field may be enhanced by microencapsulation of the BCA in an inert polymer matrix and the slow release of phage from this matrix.

Fungal viruses (mycoviruses) have also been used as BCAs. Mycoviruses are present in all major taxa of fungi. They do not appear to have mechanisms of tissue infection but rather are transmitted by hyphal anastomosis, and thus can only be exchanged between vegetatively compatible strains. In the majority of cases infection does not appear to cause any symptoms although in some cases mycovirus infection results in a hypovirulent phenotype. The most famous example is the Chestnut Blight pathogen Cryphonectria parasitica which has devastated chestnut populations in the USA and Europe. Application of virus infected hypovirulent strains to chestnut trees resulted in transmission of the virus to virulent strains by hyphal anastomosis with attenuation of virulence and protection of the trees. Whilst this strategy was successful in Europe, it did not work in the USA because of vegetative incompatibility between the strains prevented transmission of the virus to the pathogenic strains. Hypovirulence inducing mycoviruses with the ability to infect host fungal tissue when applied externally without the need for anastomosis have been identified in the fungal pathogens Sclerotinia sclerotiorum and Rosellinia necatrix. These are likely to be particularly useful as BCAs as their spread will not be limited by vegetative incompatibility.

Protection of Plants by Endophytes

Stimulation of Plant Growth

A common effect of the application of a rhizospheric or endophyti BCA to a plant is accelerated growth of the plant. Many bacterial and fungal BCAs produce analogues of plant growth regulatory hormones and volatile compounds that stimulate growth. The growth increase can be quite substantial. In one experiment inoculation of lettuce with growth promoting strains of Bacillus resulted in a 30% increase in plant weight two weeks after inoculation. Thus besides disease suppression, another advantage of biocontrol is increased yield even in the absence of disease. Volatiles such as 2,3- butanediol, acetoin, and aldehydes and ketones are produced by bacteria and may play a part in promoting plant growth. In activation of genes for synthesis of the volatiles 2,3-butanediol and acetoin in the B. subtilis biocontrol strains BSIP1173 and BSIP1174 disrupted stimulation of the host plant growth. Fungal BCAs also stimulate growth of the host plant. Trichoderma harzianum produces a butenolide metabolite called harzianolide that both stimulates growth and induces defence mechanisms.

Analogs of plant hormones produced by endophytic bacteria not only promote growth of the plants but they alleviate other stresses such as drought. For example, abscisic acid and gibberellins produced by the bacterial endophyte Azospirillum lipoferum have been shown to be involved in alleviating drought stress symptoms in maize.

Induction of Host Defence Mechanisms

Another mechanism commonly associated with protection of plants by BCAs is induction of the host defence pathways. This occurs as a result of the release of elicitors (proteins, antibiotics and volatiles) by the BCA that induce expression of the genes of the salicyclic acid pathway or the jasmonic acid/ethylene pathway. A different defence mechanism, Induced Systemic Resistance (ISR), characterised by broad spectrum resistance against pathogens of various types as well as abiotic stresses is also induced. Induction of ISR usually involves a primed state for an enhanced reaction to a biotic or abiotic stimulus rather than full induction. Because this is not full induction it is considered to require less energy than full induction and consequently have less of a negative impact on growth. Bacterial volatiles have also been implicated in induction of systemic resistance in the host plant via an ethylene dependent pathway. In addition to volatiles ISR is induced by siderophores and cyclic lipopeptide antibiotics.

Secretion of Polysaccharide Degrading Enzymes

Secretion of a variety of polysaccharide degrading enzymes including chitinases, glucanases, proteases and cellulases is a common feature of bacterial and fungal BCAs. These enzymes are capable of degrading the cell walls of fungal (or oomycete) hyphae, chlamydospores, oospores, conidia, sporangia, and zoospores resulting in lysis and thus contribute to the protection of the plant. The oligosaccharides released from degradation of the fungal cell walls act as signaling molecules to induce the host defence mechanisms. However the production of enzymes capable of degrading the hyphal cell walls of pathogenic fungi in vitro does not constitute proof that these enzymes are responsible for biocontrol activity in planta. Michelsen and Stougaard showed that although Pseudomonas fluorescens In 5 produces chitinase and beta-1,3-glucanase the biocontrol activity exhibited by this strain is not due to these enzymes but to the production of the non-ribosomal peptide antibiotics nunamycin and nunapeptin. In other studies Kim et al. found that bacterial chitinase production is not responsible for biocontrol of phytophthora blight of pepper, whilst Worasatit et al.showed that there was no relationship between the biocontrol activity of Trichoderma koningii and the production of chitinase, glucanase, or cellulase by the fungus. However, contrasting results were provided by Chernin et al. who showed by gene inactivation that chitinase production is responsible for biocontrol activity of Enterobacter agglomerans, and by Downing and Thomson who transformed a Pseudomonas strain with a chitinase gene thus creating a BCA.

Production of Antibiotics

Many biocontrol bacteria and fungi produce multiple antibiotics (including biosurfactants with antibiotic properties such as lipopeptides) that confer a competitive advantage by eliminating other bacteria and fungi. Single strains can produce multiple variants of each type (reviewed in Raaijmakers et al. and Jan et al. In addition to their antibiotic properties, lipopeptides are important signalling molecules and affect processes such as motility, induction of host plant defence mechanisms, and formation of microbial biofilms on the inner and outer surfaces of plants. The fungus Trichoderma which is widely used as a biocontrol agent and which forms the basis of several commercial products for biocontrol also synthesizes an array of secondary metabolites with antibiotic activity. Among these are the nonribosomal peptides which form voltage dependent ion channels in membranes; polyketides of unknown function; isoprenoid derivatives that are highly fungitoxic and phytotoxic; and pyrones with antifungal activity.

Various studies have attempted to provide evidence for a role for these antibiotics in pathogen suppression by enhancing their synthesis or disrupting the genes for their synthesis. Inactivation of antibiotic synthesis genes in various species of Pseudomonas, or Bacillus has provided strong evidence for the role of antibiotics in biocontrol by these species. Initial observations showed that a tryptophan auxotrophic mutant of P. aeruginosa deficient in phenazine synthesis was ineffective at supressing infection of cocoyam by Pythium myriotylum in contrast to the wild type strain which effectively suppressed infection. Disruption of rhamnolipid and phenazine synthesis genes in the species Pseudomonas aeruginosa and Pseudomonas chlororaphis significantly reduced the ability of this species to suppress Verticillium microsclerotia. However it did not completely remove the suppression suggesting that there are other mechanisms of pathogen suppression. Subsequent experiments in which the darA and darB genes responsible for the biosynthesis of the antibiotic 2-hexyl, 5-propyl resorcinol (HPR) in Pseudomonas chlororaphis were inactivated confirmed the role of the antibiotic in antagonism. Similarly, gene disruption was used to provide evidence for roles for fengycin and iturin in biocontrol of peach and curcubit diseases respectively by strains of Bacillus subtilis and of iturin in biocontrol of fruit diseases by Bacillus amyloliquefaciens. More recent work suggests that different antibiotics from the same strain interact synergistically to achieve disease suppression. A Pseudomonas strain producing phenazine and two types of cyclic lipopeptide antibiotics (sessilins and orfamides) suppressed infection of Chinese cabbage by R. solani AG2-1. Although production of phenazine alone was sufficient to achieve disease suppression, in the absence of phenazine both sessilins and orfamides were required. In suppression of root rot of bean caused by R. solani 4-HG1 both phenazines and either sessilins or orfamides were required. This study also demonstrates that the lack of an effect upon inactivation of the synthesis of a single antibiotic in a biocontrol strain does not preclude a role for that antibiotic in biocontrol.

Despite the evidence produced by the above studies showing that antibiosis is the basis for biocontrol activity in a number of species, a number of studies have produced conflicting results. Poritsanos et al.reported that a GacS mutant of P. chlororaphis was greatly reduced in the production of phenazine and showed ten fold reduction in biocontrol efficacy. However, the GacS mutation also affected the production of protease, lipase, and biofilm formation all of which would contribute to biocontrol efficacy. The biocontrol yeast Pseudozyma flocculosa (syn: Sporothrix flocculosa) is an effective biocontrol agent for control of powdery mildew. The yeast produces a powerful antibiotic that induces a rapid cell collapse in the pathogen. Despite the initial indications that the antibiotic is responsible for the biocontrol activity, it turned out not to be involved. In contrast to experiments showing that disruption of antibiotic synthesis genes in species of Pseudomonas and Bacillus reduced biocontrol efficacy, Mazzola et al. found that disruption of synthesis of the cyclic lipopeptide antibiotic, massetolide in Pseudomonas fluorescens by Tn5 insertion did not affect biocontrol activity. The demonstration of the involvement of antibiosis as amechanism of biocontrol is complicated by the plethora of antibiotics produced by individual bacterial strains. In addition, many antibiotic synthesis genes are only synthesized at high cell density, or when the bacterium forms part of a biofilm. Many such cryptic genes have been detected in the genomes of filamentous fungi, in particular Aspergillus spp. and actinomycetes. Demonstrating the involvement of antibiosis in biocontrol is further complicated by the fact that antibiotics often have additional roles other than inhibiting the growth of microorganisms. Surfactins for example are important in motility of cells on the plant surfaces, in triggering the formation of biofilms and induction of host defences. Thus inactivation of cyclic lipopeptide antibiotic genes leads not only to decreased antibiosis, and impaired host induction but also decreased ability to form biofilms. Thus antibiotics act in multiple ways to suppress pathogens.

Biofilms

On plant surfaces bacteria rarely exist as single cells, but form large multicellular assemblages called biofilms. Biofilms typically contain multiple bacterial, or mixed bacterial and fungal species. In a biofilm the cells are covered by a matrix that protects them from desiccation, UV, predation, and bactericidal compounds such as antibiotics. The matrix consists of soluble gel forming poly-saccharides, protein, lipid and DNA as well as insoluble amyloids, fimbriae, pili and flagella and is permeated by channels that act as a circulatory system for exchange of nutrients, water, enzymes, signalling molecules and removal of toxic metabolites. Biofilms are complex sorbent systems with both anionic and cationic exchangers, which means that a very wide range of substances can be trapped and accumulated for possible consumption by cells in the biofilm. The nutrient capture efficiency of the matrix exceeds that of free living cells. Not only nutrients, but also toxic substances, can accumulate in biofilms by binding to the matrix. In this way the matrix soaks up toxic substances that would otherwise be inhibitory to the cells. These substances are either retained in the matrix until it decomposes, or released from the matrix into the water phase and exuded from the matrix. Biofilm formation also facilitates the exchange of genetic information between cells. Conjugation has been shown to be 700 times more efficient in biofilms than in free-living cells. Biofilms aid in plant protection by preventing access to the surface of the plant by the pathogen, and by the production of antibiotics, many of which are only produced when growing in a biofilm. Just as biofilms may aid the survival and proliferation of biocontrol species on plant surfaces, so may they aid the survival and proliferation of pathogenic species. Additionally, cell wall degrading enzymes secreted by the pathogen bind to the biofilm matrix leading to increased heat tolerance and protection against enzymatic degradation.

Biofilm formation initiates with the aggregation of cells on the plant surface a process that is triggered by the secretion of AHLS signaling molecules by neighbouring cells. The aggregation is facilitated by components such as surfactin which modify the surface properties to enhance motility and adhesins. Once aggregation has initiated the cells synthesize the components for the matrix.

Competition for Nutrients

Competition for nutrients on, or proximal to the plant surface (rhizosphere) is another mechanism used to protect plants from pathogens. BCAs compete for sugars on the leaf surfaces or root exudates in the rhizosphere. These same food sources are required for initial establishment of the pathogen prior to infection. By utilising these food sources the BCA prevents establishment of the pathogen. For these reasons hypovirulent variants of the pathogen make effective BCAs. They occupy the same niches as the pathogen, utilise the same nutrients, and can occupy entry points to the plant tissues that would be used by the pathogen thereby preventing infection by the pathogen. Biocontrol species are able to sequester iron for their own use by the production of iron binding siderophores. This reduces the availability of iron to other organisms such as pathogens. Because bacterial siderophores have a higher affinity for iron than fungal siderophores, they are effective at depriving fungi of iron.

Inactivation of Pathogen Phytotoxins

Many plant pathogens produce phytotoxins that contribute to pathogenicity by disrupting process in the host plant. These toxins either act as enzyme inhibitors (HC toxin of Helminthosporium carbonum), interfere with membrane function (syringomycin of P. syringae), or prevent induction of host defences (coronatine of P. syringae). BCAs can protect plants from phytotoxins by inactivating them or preventing their production. The potent BCA Burhholderia heleia PAK1–2 prevents synthesis of the phytotoxin tropolone by the rice pathogen Burkholderia plantarii. Abiocontrol strain of Bacillus mycoides inactivates the toxins thaxtomin A(1) and B(2) produced by the potato common scab pathogen Streptomyces scabei. The rice sheath blight pathogen R. solani produces a host specific toxin, the RS toxin that is part of it's pathogenicity. Known biocontrol strains of T. viridae that produce an alphaglucosidase that inactivates the toxin have been isolated. The alpha glucosidase is different from other known alpha-glucosidases and is specific for the toxin. Strains of Fusarium and Trichoderma capable of inactivating the toxins Eutypine, 4-hydroxybenzaldehyde, and 3-phenyllactic acid produced by the pathogens causing Eutypa dieback and esca disease, two trunk diseases of grapevine (Vitis vinifera) have been isolated.

Genetically Modified BCAs

Techniques for genetic engineering of all organisms have been developed to a high degree of precision and have been applied to the improvement of strains of bacteria, and fungi for industrial processes. These techniques can be applied to improve the efficacy of BCAs. In one experiment the transfer of a chitinase gene from Serratia to a Pseudomonas endophyte created a strain with a greatly increased ability to suppress R. solani infection of bean. Similarly the addition of a glucanase gene to Trichoderma resulted in a strain that secreted a mixture of glucanases and showed greatly enhanced protection against the pathogens Pythium, Rhizoctonia, and Rhizopus. Zhou et al. assembled a 2,4-diacetylphloroglucinol (2,4-DAPG) biosynthesis locus phlACBDE cloned from strain CPF-10 into a mini-Tn5 transposon and introduced into the chromosome of the non 2,4-DAPG producing strain P fluorescens P32. The resultant strains provided significantly better protection of wheat against take-all caused by Gaumannomyces graminis var. tritici and tomato against bacterial wilt caused by Ralstonia solanacearum. In spite of the results of these studies these newly created BCAs are subject to the regulations that govern the use of organisms that are genetically modified through the use of recombinant DNA. Given the stiff opposition that has faced the use of transgenic plants and the even greater difficulties of containment faced with genetically modified microorganisms it is unlikely that BCAs created by recombinant DNA technology will be approved for general use in the near future.

A more realistic approach would be to use nonrecombinant DNA technology to enhance BCAs. Clermont et al. used genome shuffling to generate improved biocontrol strains of Streptomyces melanosporofaciens EF- 76. Two rounds of genome shuffling resulted in the isolation of four strains with increased antagonistic activity against the potato pathogens Streptomyces scabies and Phytophthora infestans. Chemical mutagenesis has been used to enhance biocontrol activity, e.g. nitrosoguanidine mutagenesis of Pseudomonas aurantiaca B-162 resulted in the isolation of a strain with threefold elevated levels of phenazine production and enhanced biocontrol activity. Marzano et al. isolated strains of T. harzianum with greatly enhanced biocontrol activity after UV mutagenesis. Because the genetic techniques used in these studies do not involve recombinant DNA, they simply mimic what happens naturally they do not fall under the regulations governing the use of genetically modified organisms and hence they should be more acceptable to being used for disease control. However one of the potential problems with such agents is that aside from the desired mutation there may be additional mutations in other genes that can result in undesirable consequences. More recently developed techniques of genome editing can overcome these limitations. Using tools such as Crispr/Cas we can with great precision introduce mutations into specific locations in the genome with great efficiency. An additional advantage is that mutations can be induced in multiple genes simultaneously and this will be an advantage is identifying the role of different genes in biocontrol.

Biological Control Agents

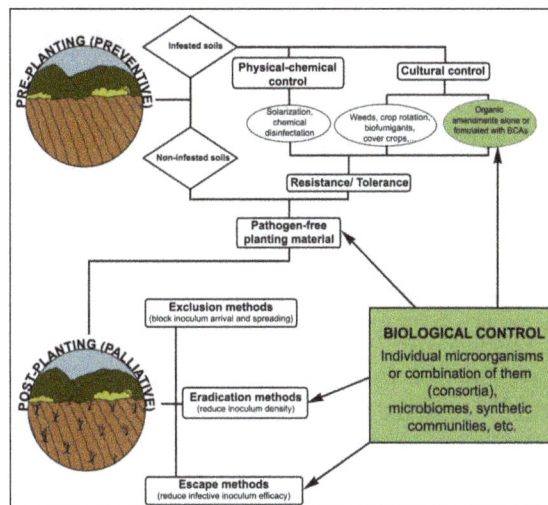

Biological control agents or Beneficials are natural enemies or competitors of crop pests. They do not harm your crop but they prey on or compete with crop pests.

Beneficials may occur naturally in and around your farm. Biological control agents are natural enemies of crop pests which have been bred in commercial insectaries and can be purchased and released into crops to control specific pests.

Predatory insects eat the eggs and/or larvae or grubs of the pest. Parasitoid insects lay eggs in the eggs or larvae of the pest. Pathogenic beneficials are diseases which affect pests.

All beneficials need to be recognised and encouraged on your farm and in your greenhouse. Beneficial organisms which occur naturally on your farm are a free method of pest control.

It is important to note that some pesticides can harm beneficials. Always look for pesticides that are compatible with biological controls before you spray.

Using biological controls requires good production and management skills. As part of an integrated pest management program, biological controls can significantly reduce the need for toxic pesticides.

1. Recognising Key Biological Controls

2. Encarsia (Encarsia Formosa): Encarsia is a tiny wasp which parasitises the eggs of greenhouse whitefly. It is available commercially and comes on small cards (from which they hatch and emerge). The cards are hung on plants throughout the greenhouse.

3. Entomopathogenic Nematode (Steinernema Feltiae): Steinernema is a nematode which attacks the larvae of fungus gnats and some caterpillars. It is available commercially as a powder that is mixed with water and added to the substrate.

4. Hypoaspis (Stratiolaelaps (Hypoaspis) miles): Hypoaspis is a predatory mite which feeds on the larvae of fungus gnats and even some western flower thrips pupae. It is available commercially in a bran mix which is added directly to the substrate.

5. Montdorensis (Typhlodromips Montdorensis): Montdorensis is a predatory mite which feeds on the larvae of western flower thrips, tomato thrips and onion thrips. It also feeds on broad mites and tomato russet mite. Montdorensis prefers warmer situations. This biological control was developed at the National Centre for Greenhouse Horticulture and is distributed onto the leaves.

6. Neoseiulus Cucumeris: This is a new beneficial for use in greenhouse cucumbers. It is a predatory mite that can be used to control thrips. It prefers cooler areas. It is purchased in vermiculite which is sprinkled onto crop foliage.

7. Occidentalis (Typhlodromus Occidentalis): Occidentalis is a predatory mite which feeds on two-spotted mites.

8. Persimilis (Phytoseiulus Persimilis): Persimilis is a predatory mite which feeds on two-spotted mites and bean spider mite. It is available commercially and comes in a small plastic bucket on bean leaves that are distributed into the crop foliage.

9. Trichogramma (Trichogramma Pretiosum): Trichogramma is a tiny wasp which attacks some species of caterpillar. It is purchased inside small cards which are placed in the crop. The wasp emerges from the cards.

10. Green Lacewing (Mallada Signatus): The green lacewing is a general predator insect and its larvae feeds on soft bodied insects such as aphids and mealy bugs. Although available commercially, it does not reproduce in greenhouses.

11. Aphidius (Aphidius Colemani): Aphidius is a small wasp which is a parasite of many aphid species. It lays an egg into the body of aphids which then hatches and feeds on the aphid before it emerges to find more aphids. It is a new bio-control agent for greenhouses and comes in small vials that contain about 500 mummified aphids with young Aphidius inside them.

12. Biorationals: Biorationals are substances used to control pests (or diseases) which have very limited or no affect on non-target organisms. These pesticides are safe to the user and are compatible with biological controls due to their selective nature or short residual activity. Biorationals are almost the ideal pesticide.

Biorational products include oils, soaps and insect growth regulators. Many of the new chemistries as well as microbially-derived products and living microbes such as fungi and bacteria are classed as biorationals.

There is on-going research in Australia including at the NCGH to identify and develop biorational pesticides.

Fungi – As Agents of Biological Control

The use of fungi in biocontrol is not a new idea. The concept was developed in the late 1800's to early 1900's. Some attempts were made to utilize fungi in controlling insects and weeds, but with the development of insecticides and weed killers, the search for biocontrol agents waned since chemical sprays were relatively cheap and efficient. However, in the last several decades, problems such as carcinogenic compounds in many sprays, gradual resistance, in the case of insects and their slow biodegradability in the environment have brought about a renaissance in biocontrol agents.

Common sense should tell us that biocontrol is not going to be completely effective. As the term implies, this is a means of "controlling" the problem and does not eradicate the target organism. Since the organisms that we are using to control another organism would normally have encountered each other, it can only reduce the population size of the pest organism to an acceptable level. If it was possible for a biocontrol agent to completely eliminate an unwanted organism, it would most likely have driven it to extinction long ago. Thus, it is best to expose a biological control agent that has not been encountered by the pest organisms.

Bioinsecticides

The use of fungi in controlling insect is somewhat unique because we have a case here where the insects now don't have to ingest anything in order for them to be eliminated. The fungi have the ability to penetrate the insect cuticle. Although there are many species of fungi that are pathogenic to insects, only a handful have been studied, only six have been registered for pest control. However, before fungi can be utilized on a large scale as insect controlling agents, a great deal of research must be carried out.

The number of species of fungi, pathogenic to insects, is not only large, but also represented in almost all taxonomic categories with the exception of some Basidiomycota and Deuteromycota. The virulence of these species also are quite variable, ranging from species which are obligate pathogens, to those that attack only weakened hosts, to some that are apparently commensal to symbiotic fungi. The species of fungi that have been targeted are those that are obligate parasites. One reason for this is host specificity. It would not do well to find a virulent pathogen that had the possibility of shifting to a different host, especially if that host would be of value, economically speaking. Also, flagellate fungi have been examined for possible use for insects that spend part of their life cycle in an aquatic environment. For example, the mosquito larval stage is found only in an aquatic environment.

Insect Biological Control Fungi

Most species of fungi that have been studied belong to the Entomophthorales (Zygomycota). This order consists of a large number of species of fungi that are parasitic on insect. In utilizing fungi as biological control agents against insects, the following categories of treatment are recognized:

- Permanent introduction of a fungus to an area with a host population. This would involve establishing the fungus species at the site of the host population. This makes a great deal of sense since if the fungus was native to that area, the insect would either have adapted to the fungal parasite or that particular insect would not be a problem in that area because the fungus would have wiped it out. An introduction would be effective since the introduction of the fungus would expose the insect to a pathogen which it had not previously encountered. This method is one of the least costly and labor intensive methods, involving periodic release of fungal spores, in order to maintain a high density of the biocontrol fungus.

- Inoculative augmentation involves releasing the pathogen in the field where it will control the insect pathogen. However, the inoculation of the fungus is not expected to carry on over to the following year. This method is used for those diseases that are an annual problem of crops.

- Conservation or Environmental Manipulations involves the modification of the host environment to enhance the probability that the fungus will infect and eliminate the host. For example, the spraying of a mild chemical insecticide that would serve to only weaken the host, making it more likely that the fungus will infect and eliminate the host. Another common means by which the environment might be modified would be to maintain a constantly wet environment in order to favor fungal growth.

The means by which the hosts are infected is usually by spores, specifically conidia since they are produced in such large number than sexual spores and are continually produced by fungi as long as conditions for growth remain favorable. At this time, there is still not a great deal of research carried out with regards to optimizing the production of conidia in fungi. Most of the technology involved in mass producing fungi has been for the purpose of extracting metabolites produced by the fungi. Thus, in some cases, the mycelium is grown, in large fermentation vats, similar to those fungi grown for their metabolites. However, the mycelium produced is broken into pieces and lyophilized. The broken mycelium is then dispersed into the host environment where, upon rehydration, will grow and produce copious amounts of conidia.

There is also the prospect that since the fungi kill the insects after penetrating their cuticle, it is thought that perhaps these fungi also produce metabolites that have insecticidal properties. However, even less research has been carried out in this area.

Some Specific Examples of Biological Control

Permanent Introductions: Control of Lymantria dispar, the Gypsy Moth

Entomophaga maimaiga, a species in the order Entomophthorales has been somewhat successful against the Gypsy Moth, or at least this is believed to be the case. The Gypsy Moth was introduced into Massachusetts, from France, in 1869, by Leopold Trouvelot in order to breed a better silkworm. However, the Gypsy Moth caterpillar soon escaped from his home and by 1889, the first major damage to trees occurred. With their voracious appetites, the Gypsy Moth can denude entire forests of all their leaves. Since the initial outbreak, the Gypsy Moth has spread approximately 15 miles each year.

Inoculative Augmentation: Control of Pine Moth with Beauveria bassiana

The use of Beauveria bassiana in controlling species of Dendrolimus, the Pine Moth, in the People's Republic of China, probably represents the largest program of biocontrol. At least one million hectares (1 hectare=2.47 acres) of pine forests are involved. The fungus is locally propagated, cheaply, on a bran or peat substrate, and is applied by air or ground equipment, as a spray or dust. Initially, during the 1970s, 'mortar bombs' containing firecrackers were used dispersal of the fungus in tall pine plantations to control the moth, which proved to be effective. However, this technique was abandoned in the 1980s because the price of firecrackers made it too expensive to use and the regulation of such goods as firecrackers and fireworks became stricter. It was also a potentially dangerous means of dispersing the fungus even though there were no reports of accidents involving this method. Applications are usually only needed at 3 year intervals.

Environmental Manipulation: Control of Hypera Postica, the Alfalfa Weevil with Erynia sp.

Medicago sativa, alfalfa is mostly used for forage. Alfalfa fields often have a number of common pathogens, among them is the Alfalfa Weevil. Introduction of various species of Erynia (Entomophthorales) has led to significant control of this insect pest. By cutting alfalfa early and leaving it in piles or clumps, for several days, provides a moist and warm microclimate which encourages the development of Erynia sp. This, in addition to possible light spraying with insecticide during the early growing season has been projected to produce a significant net profit for Alfalfa growers.

Other Potential Fungi as Biocontrol Agents

The genus Cordyceps is a well known group of fungi that are usually parasitic to insects and arachnids. They are widespread, but apparently have not been tested as biocontrol agents. Cordyceps militaris, the Caterpillar Fungus is probably the most common species in North America. This genus does not occur in Hawai'i. Cordyceps sinensis, which occurs in China, is used as a herbal remedy.

Coelomomyces is a genus of aquatic fungi that produce flagellated spores. The genus is mostly parasitic on mosquito larvae. The genus was first discovered in 1921, but as was the case with many aquatic fungi, the life cycle was incompletely known at that time. It was not until 1974 that Howard Whisler was able to observe the entire life cycle. It was only through his efforts that Coelomomyces was found to require a second host in order to complete its life cycle. This then is only the second group of fungi in which a second host is required to complete its life cycle (many species of rust, you should recall commonly require a second host). The second host is a copepod. With this discovery, it was now possible to induce the fungus to produce spores in the laboratory that can be used to inoculate mosquito-infested areas. Such a project would probably be useful, in Hawai'i, where mosquitoes are present throughout the year. However, there are problems associated with using this fungus.

Major Advantages of using Fungi for Biocontrol

Some of the obvious advantages of using fungi in biocontrol is the of the ubiquitous nature of fungi and their genetic diversity. This latter characteristic can provide a number of biocontrol agent from single species of fungi. Although there are other biocontrol agents, such as bacteria and viruses, there are some distinct advantages in utilizing fungi. Fungi have the ability to directly infect the host insect by penetrating the cuticle of the insects, something viruses and bacteria are unable to do. Instead these latter organisms must be ingestion in order to be effective as biocontrol agents. This is not possible with many insects, such as aphids, which feed exclusively by inserting their stylets into herbaceous plant organs in order to obtain plant sap. Also, there are few viral and bacterial diseases for a number of insects.

Finally, as mentioned at the beginning, fungi, in contrast to insecticides, do not pose a health hazard to people and domestic animals. Many of the more lethal insecticides have been banned for this reason. However, it should be pointed out that most fungi as biocontrol agents have, and will probably continue to have only limited success. One of the major problems that have caused negative reactions to fungi, as agents of biological control, is the overenthusiastic promises that some researchers have made in the past.

Bioherbicides

The idea of using biocontrol agents to control weeds is not a new one. The utilization of insects for this purpose have been attempted for quite sometime. The utilization of fungi for this purpose, on the other hand, has only been intensively studied during the last 20 years. Neither one, however, has really challenged the use of chemical herbicides which has been a thriving area of research, Weed Science.

In Hawaii, there are presently, several researchers that are using fungi as biocontrol agents against weeds: Dr. Donald Gardner, Plant Pathologist with the National Parks Service and Dr. Eduardo Trujillo, Professor of Plant Pathology, University of Hawaii. Dr. Gardner has more recently began carrying out research in this area and has not yet produced any significant results at this time. He has been working on biocontrol of such weeds as Passiflora tripartita, Banana Polka, utilizing a species of Fusarium as the biocontrol agent. However, Dr. Trujillo has been doing this research much longer and has a certain amount of success. Among some of the weeds that he has worked on include Ageratina riparia, Hamakua Pamakani, which has been controlled in some areas in Hawaii by the smut fungus, Entyloma compositarum, which was introduced from Jamaica. Because of the strict quarantine laws here in Hawaii, exhaustive tests had to be carried out in order to demonstrate that the pathogen would not spread to other species of plants. This was done. That's another advantage that the use of fungi, as biocontrol agents, have. There are numerous species that are very specific as to what host they will attack. Release of this fungus was very successful in some areas. Under ideal conditions, i.e. high rainfall and temperatures between 10-18 °C, the fungus would devastate the host plant.

One of Dr. Trujillo most successful biocontrol species is Colletotrichum gloeosporioides f.sp. clidemiae which was introduced from Panama. This species was introduced to control the noxious weed, Clidemia hirta, Koster's curse. Colletotrichum was very successful in devastating Clidemia in areas where it was introduced. However, it appears to have a poor dispersal mechanism and the control of Clidemia is usually pretty localized. Other species that Dr. Trujillo has attempted to control include Lantana camara, an ornamental shrub that has escaped cultivation and also Banana Polka since the retirement of Dr. Gardner.

Not all species of fungi used as biocontrol were introduced from elsewhere. An unidentified species of Cephalosporium was isolated from a Kauai ranch in 1968, and caused a wilt on Cassia surattensis, Kolomona. Application of the fungus is by spore suspensions in water which is sprayed on man-made wounds on trunks of healthy trees.

Commercial Mycoherbicides and Mycoinsecticides

There are presently only a few mycoherbicides and mycoinsecticides that are available and they have had limited success. However, a number of mycofungicides are available commericially. A list of products are given below.

AQ10	RootShield
Aspire	SoilGard (formerly GlioGard)
Binab T	Trichodex
Biofox C	Trichodowels
Bio-Fungus (formerly Anti-Fungus)	Trichopel
Contans	Trichosea
Fusaclean	
Rotstop, P.g. Suspension	

The most common species of fungi used in this type of biocontrol is Trichoderma harzianum. Other species of Trichoderma also have been used, as well as other genera of fungi.

This species was genetically engineered, at the Cornell Plant Pathology Department and is known to infect a number of different species of fungal plant pathogens, and can be used selectively on different parts of the plant body. In the product RootShield, it is disseminated in the soil as a wet able powder or as granules and can be applied with fertilizer, when fertilizing plants. Once released the fungus grows around the root system and protects it from various root diseases by parasitizing and killing root pathogenic fungi. However, it is compatible with mycorrhizal fungi and chemical fungicides.

Recently, tests have been carried out to study the use of honey bees in disseminating T. harzianum. Species of Botrytis are known to cause fruit rot in strawberries, as well as in grape plants. Previously, the fungus was sprayed on the flower and fruit by suspending the spores in water and spraying the plants. However, in using honey bees, a dual function is served. Not only are the honey bees disseminating the fungus, but in visiting the flowers, the bees also pollinate the flowers. It was concluded from this test that the strawberry plants produced more strawberry due to pollination by bees and there was significantly less fruit rot due to the biocontrol fungus.

Trichoderma

Trichoderma is a very effective biological mean for plant disease management especially the soil born. It is a free-living fungus which is common in soil and root ecosystems. It is highly interactive in root, soil and foliar environments. It reduces growth, survival or infections caused by pathogens by different mechanisms like competition, antibiosis, mycoparasitism, hyphal interactions, and enzyme secretion.

General Characteristics

Colonies, at first transparent on media such as cornmeal dextrose agar (CMD) or white on richer media such as potato dextrose agar (PDA). Mycelium typically not obvious on CMD, conidia typically forming within one week in compact or loose tufts in shades of green or yellow or less frequently white. Yellow pigment may be secreted into the agar, especially on PDA. A characteristic sweet or 'coconut' odor is produced by some species.

Conidiophores are highly branched and thus difficult to define or measure, loosely or compactly tufted, often formed in distinct concentric rings or borne along the scant aerial hyphae. Main branches of the conidiophores produce lateral side branches that may be paired or not, the longest branches distant from the tip and often phialides arising directly from the main axis near the tip. The branches may rebranch, with the secondary branches often paired and longest secondary branches being closest to the main axis. All primary and secondary branches arise at or near 90° with respect to the main axis. The typical Trichoderma conidiophores with paired branches assumes a pyramidal aspect.

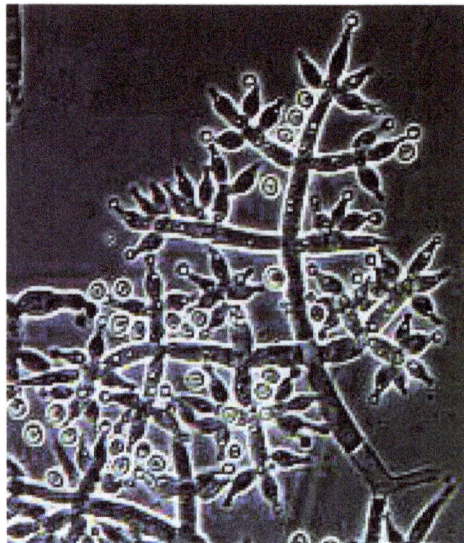

Trichoderma harzianum

Phialides are typically enlarged in the middle but may be cylindrical or nearly subglobose. Phialides may be held in whorls, at an angle of 90° with respect to other members of the whorl, or they may be variously penicillate (gliocladium-like). Phialides may be densely clustered on wide main axis (e.g. T. polysporum, T. hamatum) or they may be solitary (e.g. T. longibrachiatum).

Conidia typically appear dry but in some species they may be held in drops of clear green or yellow liquid (e.g. T. virens, T. flavofuscum). Conidia of most species are ellipsoidal, 3-5 x 2-4 μm. Conidia are typically smooth but tuberculate to finely warted conidia are known in a few species.

Synanamorphs are formed by some species that also have typical Trichoderma pustules. Synana-morphs are recognized by their solitary conidiophores that are verticillately branched and that bear conidia in a drop of clear green liquid at the tip of each phialide.

Chlamydospores may be produced by all species, but not all species produce chlamydospores on CMD at 20° C within 10 days. Chlamydospores are typically unicellular subglobose and terminate short hyphae; they may also be formed within hyphal cells. Chlamydospores of some species are multicellular (e.g. T. stromaticum).

Teleomorphs of Trichoderma are species of the ascomycete genus Hypocrea Fr. These are char-acterized by the formation of fleshy, stromata in shades of light or dark brown, yellow or orange. Typically the stroma is discoidal to pulvinate and limited in extent but stromata of some species are effused, sometimes covering extensive areas. Stromata of some species (Podostroma) are clav-ate or turbinate. Perithecia are completely immersed. Ascospores are bicellular but disarticulate at the septum early in development into 16 part-ascospores so that the ascus appears to contain 16 ascospores. Ascospores are hyaline or green and typically spinulose. More than 200 species of Hypocrea have been described but only few have been grown in pure culture and fewer have been redescribed in modern terms.

Benefits of Trichoderma

1. Disease Control: Trichoderma is a potent biocontrol agent and used extensively for soil born diseases. It has been used successfully against pathogenic fungi belonging to various genera, viz. Fusarium, Phytopthara, Scelerotia etc.

2. Plant Growth Promoter: Trichoderma strains solubilize phosphates and micronutrients. The application of Trichoderma strains with plants increases the number of deep roots, thereby increasing the plant's ability to resist drought.

3. Biochemical Elicitors of Disease: Trichoderma strains are known to induce resistance in plants. Three classes of compounds that are produced by Trichoderma and induce resis-tance in plants are now known. These compounds induce ethylene production, hypersensi-tive responses and other defense related reactions in plant cultivars.

4. Transgenic Plants: Introduction of endochitinase gene from Trichoderma into plants such as tobacco and potato plants has increased their resistance to fungal growth. Selected transgenic lines are highly tolerant to foliar pathogens such as Alternaria alternata, A. so-lani, and Botrytis cirerea as well as to the soil-borne pathogen, Rhizectonia spp.

5. Bioremediation: Trichoderma strains play an important role in the bioremediation of soil that are contaminated with pesticides and herbicides. They have the ability to degrade a wide range of insecticides: organochlorines, organophosphates and carbonates.

Biocontrol Mechanisms of Trichoderma

The Trichoderma may suppress the growth of the pathogen population in the rhizosphere through competition and thus reduce disease development. It produces antibiotics and toxins such as trichothecin and a sesquiterpine, Trichodermin, which have a direct effect on other organisms.

The antagonist (Trichoderma) hyphae either grow along the host hyphae or coil around it and secrete different lytic enzymes such as chitinase, glucanase and pectinase that are involved in the process of mycoparasitism. Examples of such interactions are T. harzianum acting against Fusarium oxyporum, F. roseum, F. solani, Phytophthara colocaciae and Sclerotium rolfsii. In addition, Trichoderma Enhances yield along with quality of produce. Boost germination rate. Increase in shoot and Root length Solubilizing various insoluble forms of Phosphates Augment Nitrogen fixing. Promote healthy growth in early stages of crop. Increase Dry matter Production substantially. Provide natural long term immunity to crops and soil.

Method of Application

1. Seed treatment: Mix 6 - 10 g of Trichoderma powder per Kg of seed before sowing.

2. Nursery treatment: Apply 10 - 25 g of Trichoderma powder per 100 m2 of nursery bed. Application of neem cake and FYM before treatment increases the efficacy.

3. Cutting and seedling root dip: Mix 10g of Trichoderma powder along with 100g of well rotten FYM per liter of water and dip the cuttings and seedlings for 10 minutes before planting.

4. Soil treatment: Apply 5 Kg of Trichoderma powder per hector after turning of sun hemp or dhainch into the soil for green manuring. Or Mix 1kg of Trichoderma formulation in 100 kg of farmyard manure and cover it for 7 days with polythene. Sprinkle the heap with water intermittently. Turn the mixture in every 3-4 days interval and then broadcast in the field.

5. Plant Treatment: Drench the soil near stem region with 10g Trichoderma powder mixed in a liter of water.

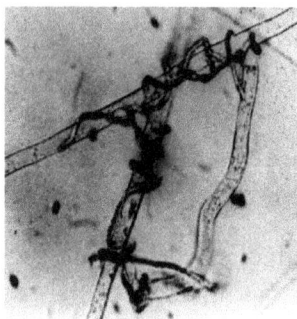

Trichoderma's mycelium parasitising to mycelium of Pythium

Trichoderma Formulations

Important commercial formulations are available in the name of Sanjibani, Guard, Niprot and Bioderma. These formulations contain 3 x 106 cfu per 1 g of carrier material. Talc is used as carrier for making powder formulation.

Uses

Used in Damping off caused by Pythium sp. Phytophthora sp., Root rot caused by Pellicularis filamentosa, Seedling blight caused by Pythium, Collar rot caused by Pellicularia rolfsii, Dry rot caused by Macrophomina phaseoli, Charcoal rot caused by Macrophomina phaseoli, Loose smut caused by Ustilago segetum, Karnal bunt diseases, Black scurf caused by Rhizoctonia solani, Foot rots of Pepper and betel vine and Capsule rot of several crops. Effective against silver leaf on plum, peach and nectarine, Dutch elm disease on elm's honey fungus (Armillaria mellea) on a range of tree species, Botrytis caused by Botrytis cinerea, Effective against rots on a wide range of crops, caused by fusarium, Rhizoctonia, and pythium, and sclerotium forming pathogens such as Sclerotinia and Sclerotium.

Trichoderma is most useful for all types of Plants and Vegetables such as cauliflower, cotton, tobacco, soybean, sugarcane, sugarbeet, eggplant, red gram, Bengal gram, banana, tomato, chillies, potato, citrus, onion, groundnut, peas, sunflower, brinjal, coffee, tea, ginger, turmeric, pepper, betel vine, cardamom etc.

Precautions

- Don't use chemical fungicide after application of Trichoderma for 4-5 days.
- Don't use trichoderma in dry soil. Moisture is a essential factor for its growth and survivability.
- Don't put the treated seeds in direct sun rays.
- Don't keep the treated FYM for longer duration.

Compatibility

Trichoderma is compatible with Organic manure Trichoderma is compatible with biofertilizers like Rhizobium, Azospirillum, Bacillus Subtilis and Phosphobacteria.

Trichoderma can be applied to seeds treated with metalaxyl or thiram but not mercurials. It can be mixed with chemical fungicides as tank mix.

Biopesticide

Biopesticides are materials with pesticidal properties that originate from natural living organisms, including microorganisms, plants, and animals. There are three major classes that biopesticides fall into:

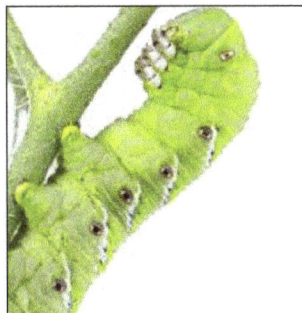

1. Microbial pesticides: These biopesticides are produced by microorganisms, including bacteria, viruses, and certain fungi. Each type of microbial pesticide targets a specific species or small group of species. It is common that microbial pesticides control a large variety of pests.

2. Biochemical/herbal pesticides: These are substances naturally occurring in the environment that control pests. This could include plant extracts that lure and trap insects or insect pheromones that interfere with mating. It may include botanical extractions that are active against plant disease pathogens and other pests.

3. Plant-Incorporated Protectants (PIPs): These pesticides are produced from plants as a result of another genetically incorporated material added to that plant (aka GM crops). While this application of pesticides originates from natural material, it also interferes with the natural biochemistry of the target organism and is thus widely contested.

As a natural pesticide choice, natural microbial and biochemical pesticides are the type commonly used by farmers and growers to control an existing pest problem, because they can be applied like synthetic pesticides but without the toxic damage.

Microbial and biochemical pesticides are increasingly used as soil amendments or seed treatments that will target the necessary area of the plant. When microorganisms are added to the soil/plant complex they release families of biochemical molecules to a targeted environment, such as the surface of the leaf or stem or in the root rhizosphere. The pesticidal properties of the microbial biochemical excretions then aid the plant in its affected areas.

The most widely used microbial pesticide are types of the bacteria Bacillus thuringiensis or Bt. Each strain of bacterium produces different proteins that are toxic to certain insects, specifically targeting insect larvae.

Naturally derived materials such as copper, baking soda, sesame oil, clove oil, rosemary oil and canola oil are also considered biopesticides.

Advantages of Biopesticides

So why use biopesticides when you can buy other synthetic pesticides anywhere? For one, biopesticides have simply been known and understood for millions of years. This means that plants are well familiar with and responsive to biopesticides, without major adverse side effects. Also, natural microbial and biochemical materials are common to our global ecology and are easily processed in our ecology with minimal probability of environmental imbalance.

This is the significant difference from synthetic pesticides. Plants and other living organisms are not accustomed or responsive to synthetic molecules, and this unfamiliarity results in rejection or negative reaction and side effects. The plant doesn't know how to store these synthetic chemicals, so its cells can become cancerous, growing abnormally and causing mistakes in normal biological function.

Biopesticides also show a number of other advantages over synthetic pesticides, including:

More renewable,

- Biodegradable.

- Can be more effective in the long-term.

- Effective in small quantities and quickly decompose, avoiding pollution, which is a major problem with synthetic pesticides.

- Can be less expensive.

- Affect only targeted pests, unlike broad spectrum synthetic pesticides that can take effect on other unintended insects, birds, and mammals, including humans.

- Have a host of natural compounds which may be active against disease and pest, thus minimizing the capacity of pathogenic and invasive organisms to adapt to the molecules, thus rendering them ineffective.

In short, biopesticides prove to be a predictable and less toxic form of pest control compared to the less predictable and more toxic conventional synthetic pesticides.

Biopesticides in Organic Farming

Organic growers have available a large array of biopesticides that may be applied for the management of plant diseases. While generally a logical fit for managing diseases in organic crops, there are some biopesticides that are not approved for organic production. Some biopesticides, like salts of phosphorous acid and all genetically-engineered PIPs, are not allowed for organic production. Note that some fungicides approved for use in organic production systems are not biopesticides, including mineral oils, copper, and sulfur. It is critical to verify that a biopesticide, like any other product, is approved for organic production and registered in your state prior to any application.

Benefits and Limitations of Biopesticides

As Benjamin Franklin once said in regards to fire prevention, "an ounce of prevention is worth a pound of cure." The same holds true for disease control, especially when using biopesticides. Products should be used in a preventative and not a curative manner, as they typically lack the breadth of activity, efficacy, or residual activity of conventional fungicides. And, it is more important to apply biopesticides to the specific targets for which they have been shown to be most effective during the interval when those pathogens are most likely to be active. Failure to match the biopesticide to the appropriate disease or timing thereof will reduce the value of such applications.

Many biopesticides function by interfering with a pathogen's ability to infect a susceptible plant either directly through parasitism or indirectly through the production of secondary compounds. Therefore, timing and frequency of application are critical. Effective control requires that the application of materials begins prior to conditions favorable for disease development or immediately following the first symptoms of disease. In foliar applications, these materials provide a protective barrier on plant surfaces that interfere with the pathogen. Repeated applications are required, often on a weekly interval, due to loss of product from degradation or rain/irrigation washing materials off foliage and fruit, or with the growth of new foliage. Some biopesticides, notably Regalia and Companion, are thought to turn on plant defense responses. Such products should be applied before infection to be effective.

Similar strategies are at play when biopesticides are applied to seeds, roots, or soil. These materials can provide temporary protection to roots, but results can vary depending on the crop, pathogen, and environment. Subsequent root growth or rain/irrigation may wash biochemical materials away from the root zone. Many biological agents can grow within the root zone and can persist longer than chemicals. However, because they are often unable to compete with native soil microorganisms, they tend to persist at levels ineffective for control. Many biopesticides are effective in the control of root rots associated with seed germination and seedling/transplant establishment. Again, the efficacy of biopesticides and most fungicides is limited against established infections, so the sooner applications are started, either prior to or during a disease outbreak, the more likely control will be realized.

The frequency of application of any product is related to its residual time, or the time required for a product to degrade in the environment. Biopesticides in general have a short residual time, typically much shorter than compounds such as copper. On the other hand, this short residual time also limits most concerns about build-up in the soil that are associated with copper. In addition, the activity of biopesticides is generally targeted to specific pests and closely related organisms, and they are usually inherently less toxic to non-target organisms. Therefore, they do not have the same potential to affect birds, beneficial insects, and mammals. Because of the short residual time, low toxicity, and reduced risk to nontarget organisms or the environment, the EPA generally requires less data to register a new biopesticide compared to a conventional pesticide. However, sufficient data regarding the composition, toxicity, degradation, and other characteristics of the pesticide is still required to ensure that the product will not have any adverse effects on human health or the environment.

The EPA, recognizing that biopesticides tend to pose fewer risks than conventional pesticides, has taken steps to promote development and adoption of new products. The registration process is quicker for biopesticides than conventional products, often taking less than a year, compared with an average of more than 3 years for conventional pesticides. To facilitate their registration, the Biopesticides and Pollution Prevention Division was established in the Office of Pesticide Programs in 1994. Some biopesticides are defined as minimum-risk pesticides through the FIFRA Section 25(b) rule because their active and inert ingredients are generally recognized as safe (GRAS). These consequently are exempted from the regulation requirements of FIFRA and thus can be used on any labeled crops for any target since they do not need to be registered as a pesticide. "Exempt from EPA registration" is stated on the label of these products.

When selecting a biopesticide it is important to obtain information about the product's mode of action, residual time, and target disease(s). Unfortunately, data on the efficacy of biopesticides is limited. Many products are broadly labeled with separate lists of registered crops and labeled diseases. Data proving efficacy is not required for the registration of pesticides in the USA. Most biopesticides are produced by small companies that lack funding to support replicated field trials needed to obtain sufficient efficacy data by experienced university and other independent researchers. To help fill this gap, the IR-4 Biopesticide and Organic Support Program funds grants to obtain efficacy information for biopesticides in development as well as those already registered. These funded projects help the program meet its objective, which is to further the development and registration of biopesticides for use in pest management systems for specialty crops (which include all vegetables) or for minor uses on major crops.

Several factors can affect performance of biopesticides, or any product used for disease control. These need to be considered when reviewing results from an efficacy experiment or on farm use. It is especially challenging to assess efficacy of a product used on-farm because there are no plants left untreated or treated with products known to be effective for comparison. A product may appear to be effective when actually conditions became unfavorable for the disease to develop or the pathogen was not present. Conversely, conditions may be so favorable for disease development that it is not possible to suppress the pathogen even with a conventional product demonstrated previously to be effective. The earlier in disease development that applications of a product are started, the more effective the product will be. Disease spots (lesions) cannot be 'cured' and once a pathogen has infected a plant it cannot be killed, even with most conventional fungicides, in contrast with insect pests which remain on plant surfaces and are accessible to treatment. Product performance can also be affected by spray coverage and frequency of application. Treatment timing is a common potential explanation for poor control with a product that has been effective in other situations. Sometimes in efficacy experiments the pathogen is introduced artificially rather than relying on natural inoculum. This may result in disease pressure that is greater or less than what would occur naturally Some products have continued to be developed and improved following registration, thus results obtained with an early formulation might not reflect the degree of control obtainable with the current formulation. Performance of some products can be improved by using an adjuvant, but on the other hand, it has been suggested that some products have been negatively affected by the adjuvant used. Laboratory testing provides an indication of product activity, but these results alone are not sufficient for predicting field efficacy because of the many environmental factors that can affect performance. Finally, most efficacy studies have not been conducted in organic systems, where healthy soils, crop rotations, and biodiverse agroecological environments may effect outcomes.

Types of Bio-Pesticides

Bio-Insecticides

Microorganisms like bacteria, fungi and viruses show greatest commercial importance as bio-control agents than the commercial synthetic pesticides. With the more progress of scientific achievement the bio-insecticide becomes popular in different countries.

Bacterial

Several bacterial pathogens of different insects are being used as insecticides. These are Bacillus, Clostridium, Pseudomonas, Enterobacter, Proteus, Serratia etc. Out of these, Bacillus thuringiensis has been used extensively. The bacterium has been shown to be successful against a wide range of more than 1 50 insects.

It shows insecticidal activity against larvae of Lepidoptera. B. thuringiensis var. israelensis (BTI) is highly active against larvae of mosquito vectors of malaria. Unlike DDT, the pathogen is environmentally safe in use and mosquito does not show any resistance against the bacterium. It also shows excellent result to control black fly — the carrier of widespread river blindness in Africa.

Four different toxins are produced by the B. thuringiensis and about 16 formulations have been prepared based on the above toxins. These are used in different countries like USA, Russia, France,

Germany etc. Some of the registered products like Sporcine, Condor, Cutlass, Thuricide, Foil etc., are commonly used.

In India, it has been found that 0.4% thuricide is more effective than malathione, endrine and DDT to control insect pests of crucifer, lac and sugarcane. In USA, different registered formulations prepared from B. thuringiensis are used to control pests of different crops like Alfalfa caterpillar (Alfalfa), Bollworm (Cotton), Cabbage worm (Cabbage and Cauliflower), Orange dog (Orange), Crape leaf folder (Grapes) etc.

With the help of recombinant DNA technology, the gene having insecticidal properties of Bacillus thuringiensis has been transferred to the crops plans like tomato in 1987, with the help of bacterium, Agrobacterium tumifaciens. Later, similar success has been achieved in different crop plants like tobacco, cotton etc. by using the similar technique. Now-a-days, Bt cotton is very popular among the farmers.

Fungal

Different entomogenous fungi have been used to protect different crops from insect damage. Species of the different genera like Aschersonia, Beauveria, Coelomomyces, Entomophthora, Hirsutella etc. are in common use. Some registered products are also produced and used by different countries like Australia, Brazil, China, France, Japan, USA, UK etc.

List of some of the products, producing fungi in parenthesis and insects on which it acts are:

   i.   Aseronija (Aschersonia aleyrodis) — Whitefly of many crops.

   ii. Boverin and Boverol (Beauveria brassiana) — Pine caterpillar, Green leaf hipper, Colorado potato beetle etc.

   iii. Mycotal and Vertalec (Verticillium lecanii) — Whitefly and Aphids of glasshouse crop.

Viral

Viruses are also very much effectively used as bioinsecticide. There are three major groups of viruses that can infect different insects.

These are:

   i.   Nuclear polyhedrosis viruses (NPV),

   ii. Granulosis viruses (GV) of Baculoviridae, and

   iii. Cytoplasmic polyhedrosis viruses (CPV) of Reoviridae.

The NPVs are effective against moths and butterflies, while CPVs are effective mainly on caterpillars. These are used in different countries like USA, UK, Canada, lapan, Germany etc.

List if some viruses, registered trade names in parenthesis and target insects are given below:

   i.   NPV (ELCAR) — Tobacco budworm and Cotton bollworm.

ii. NPV (GYPCHEK) — Gypsy moth.

iii. NPV (VfROX) — European sawfly.

iv. CPV (MATSUKEMIN) — Pine caterpillar.

v. GV (MATEX) — Insects of different food crops like Codling moths.

Bio-Nematicides

Different fungi are known to act as nematicide. Fungi of different genera like Arthrobotrys, Dactylella, Dactylaria and Monacrosporium are used to control different members of genera like Heterodera, Meloidogyne and Rotylenchulus, cause diseases of different crop plants.

The Fungi Damage Nematode in Four different Ways:

By haustoria

Fungi penetrate haustoria in the body of the nematode, digest the cell contents and draw the nutrients.

Catching by Loop

The fungal mycelium forms loops at intervals. As a nematode passes through the loop, it constricts and thus the nematode is trapped.

Production of Adhesive Hyphae

The fungal mycelium produces some adhesive branch which may stick with the body of nematode on accidental contact.

Formation of Hyphal Mesh

The mycelium forms mesh-like cobweb and is able to catch nematodes.

The other groups of fungi are commonly present in the soil and can act as nematicides. These are Verticillium chlamydosporium, Paecilomyces lilacinus and Dactylella oviparasitica — they often attack nematodes as well as their eggs.

Bioherbicides

The use of bioherbicides is another way of controlling weeds without environmental hazards posed by synthetic herbicides. Bioherbicides are made up of microorganisms (e.g. bacteria, viruses, fungi) and certain insects (e.g. parasitic wasps, painted lady butterfly) that can target very specific weeds. The microbes possess invasive genes that can attack the defense genes of the weeds, thereby killing it.

The better understanding of the genes of both microorganisms and plants has allowed scientists to isolate microbes (pathogens) whose genes match particular weeds and are effective in causing a fatal disease in those weeds. Bioherbicides deliver more of these pathogens to the fields. They are sent when the weeds are most susceptible to illness.

The genes of disease-causing pathogens are very specific. The microbe's genes give it particular techniques to overcome the unique defenses of one type of plant. They instruct the microbe to attack only the one plant species it can successfully infect. The invasion genes of the pathogen have to match the defense genes of the plant. Then the microbe knows it can successfully begin its attack on this one particular type of plant. The matching gene requirement means that a pathogen is harmless to all plants except the one weed identified by the microbe's genetic code.

This selective response makes bioherbicides very useful because they kill only certain weed plants that interfere with crop productivity without damaging the crop itself. Bioherbicides can target one weed and leave the rest of the environment unharmed.

The benefit of using bioherbicides is that it can survive in the environment long enough for the next growing season where there will be more weeds to infect. It is cheaper compared to synthetic pesticides thus could essentially reduce farming expenses if managed properly. It is not harmful to the environment compared to conventional herbicides and will not affect non-target organisms.

With the advances of genetic engineering, new generation bioherbicides are being developed that are more effective against weeds. Microorganisms are designed to effectively overcome the weed's defenses. Weeds have a waxy outer tissue coating the leaves that microorganisms have to penetrate in order to fully infect the weeds. Through biotechnology, these microorganisms will be able to produce the appropriate type and amount of enzymes to cut through the outer defenses. Streamlining of the microbe's plant host specificity will ensure that the weeds are taken out and not the crops. On the other hand, microbes can also be made to be effective against several host weeds and not only to one type of weed as this can be too expensive to produce for commercial use.

Biofungicides

The term biofungicide can have several different meanings, but it is most frequently used to refer to fungicides that contain a microorganism (usually a bacterium or fungus) as the active ingredient.

Biofungicides can control many different kinds of fungi and water molds, although each separate active ingredient controls only certain pathogens. Some also control bacterial diseases. Virtually all of the organisms used in biofungicides on the market today occur naturally in soil or on plant surfaces, and most are approved for use in organic production.

Powdery mildew in roses can be managed with foliar applications of the biofungicide Bacillus subtilis.

Advances in fermentation technology have allowed mass production of highly specialized microbes that previously could only be grown in small batches on highly specific substrates, such as on roots infected with pathogens. Consumer demand for organically certifiable pesticides and increased regulatory pressure on older synthetic pesticides, especially in Europe, has fostered increased commercial interest in the production of living organisms that can suppress or kill pathogens. For these reasons and also because as natural products biofungicides generally have few negative impacts on health and the environment, the number available will likely continue to increase.

Black spot on rose can also be managed with foliar applications of Bacillus subtilis.

Since microbial biofungicides contain living organisms, their modes of action differ from those of synthetic fungicides. Some of these mechanisms include:

- Competition: The biocontrol agent is more effective than the pathogen at gathering critical nutrients or space and, therefore, must be in place before disease onset.

- Antibiosis: The biocontrol agent produces a chemical compound of some type (antibiotic or toxin) that acts against the pathogen.

- Predation or parasitism: The biocontrol agent directly attacks the pathogen.

- Induction of host plant resistance: The biocontrol agent triggers a defensive response in the host plant that limits the ability of the pathogen to invade the plant.

Most biofungicides use one or more of the above mechanisms to target only one or a few specific pests. As such, applicators should both read the label and diagnose the problem carefully to insure that the product will be effective. Biofungicides work best when applied preventively. Application after a plant is already infected has little chance of significantly altering the course of the disease for that plant, although it may decrease the ability of the pathogen to move from that plant to other plants, especially if the pathogen has to move through the soil to do so. Thus, an application of biofungicide is not likely to cure an infected plant; but it may protect other nearby plants in the field.

Biofungicides containing Streptomyces or Trichoderma can be used to prevent infection of plants
with damping off or root or seed rot pathogens such as Pythium.

Although independent testing by university researchers and others has verified some manufacturer claims made for these products, efficacy data for many other products against pathogens on ornamentals is unavailable. For instance, as of early 2014, independent testing has not demonstrated adequate field efficacy of any biofungicide for landscape or agricultural use against Armillaria root rot (otherwise known as oak root rot). However, a number of biologicals have been found effective for control of Pythium, Phytophthora, Verticillium, and other pathogens on a variety of plant hosts. With these types of products, eradication of the pathogen is not the goal and is probably never achieved. Instead, biofungicides rely on a core tenet of the IPM philosophy: keeping pest levels below damaging thresholds and using biofungicides (when necessary) in combination with cultural practices that promote healthy plant growth.

Biofungicides cannot take the place of proper cultural care. They are a valuable tool for keeping a strong plant healthy, but they cannot forestall the inevitable. If your client's Japanese maples are routinely drowned, allowed to wilt, and then drowned again, adding a biofungicide will not prevent them from contracting Phytophthora if it is present in the soil.

If biofungicides are a useful and environmentally friendly tool in the landscape, why aren't they more widely used? One reason is that these fungicides rely on living organisms for efficacy, so they must be stored appropriately in order to retain their fungicidal properties over time. However, a more commonly cited reason is that the personal protective equipment needed to apply them is more involved than for some other compounds. Routine exposure to the proteins found in the spray mists of some biofungicides can result in the development of allergic reactions. To keep commercial applicators safe, they must wear NIOSH approved respirators when mixing, loading, or applying biofungicides in agricultural or landscape settings. This may not be immediately obvious when reading the labels, as a quick scan often only shows the following required personal protective equipment (PPE):

- Long sleeved shirt and long pants.

- Shoes plus socks.

- Waterproof gloves.

The respirator requirement is only evident when reading the text following the list. Biofungicides are safe to use as long as mixer/loaders and applicators have and use a respirator as part of their PPE. However, a NIOSH approved respirator requires proper training and fitting in order to be effective.

When used properly and with forethought, biofungicides can be an important part of an IPM program to prevent or mitigate problems with plant pathogens in the landscape. However, nothing can ultimately take the place of proper plant selection and care.

References

- Biological-control: slu.se, Retrieved 11 April, 2019

- Biological-Pest-Control: allstarce.com, Retrieved 2 July, 2019

- Landis: umn.edu, Retrieved 6 March, 2019

- Conservation-biological-control-practices, biological-control-of-pest-and-vector-insects: intechopen.com, Retrieved 16 May, 2019

- Biological-control-of-plant-diseases: researchgate.net, Retrieved 28 June, 2019

- Biological, pests,-diseases-and-disorders, greenhouse, horticulture, agriculture: dpi.nsw.gov.au, Retrieved 8 February, 2019

- Use-of-fungi-as-agents-of-biological-control, botany: sites.google.com, Retrieved 20 August, 2019

- Trichoderma-bio-control-agent-management-soil-born-diseases: agropedia.iitk.ac.in, Retrieved 23 June, 2019

- Biopesticides: soiltechcorp.com, Retrieved 30 January, 2019

- Biopesticides-for-plant-disease-management-in-organic-farming: extension.org, Retrieved 20 August, 2019

- Main-types-of-bio-pesticides, plants: biologydiscussion.com, Retrieved 1 February, 2019

- Bioherbicides, biotechinagriculture: isaaa.org, Retrieved 21 July, 2019

2

Plant Diseases and Biological Control of Plant Pathogens

Plant disease can be described as an impairment of the normal condition of the plant that affects its vital functions. The organisms which can cause such infectious diseases are known as plant pathogens such as viruses, oomycetes, bacteria, viroids, fungi, nematodes, etc. The chapter closely examines the key concepts of plant diseases and pathogens to provide an extensive understanding of the subject.

Plant Disease

Plant disease is an impairment of the normal state of a plant that interrupts or modifies its vital functions. All species of plants, wild and cultivated alike, are subject to disease. Although each species is susceptible to characteristic diseases, these are, in each case, relatively few in number. The occurrence and prevalence of plant diseases vary from season to season, depending on the presence of the pathogen, environmental conditions, and the crops and varieties grown. Some plant varieties are particularly subject to outbreaks of diseases while others are more resistant to them.

Nature and Importance of Plant Diseases

Plant diseases are known from times preceding the earliest writings. Fossil evidence indicates that plants were affected by disease 250 million years ago. Other plant disease outbreaks with similar far-reaching effects in more recent times include late blight of potato in Ireland; powdery and downy mildews of grape in France; coffee rust in Ceylon (now Sri Lanka; starting in the 1870s); Fusarium wilts of cotton and flax; southern bacterial wilt of tobacco; Sigatoka leaf spot and Panama disease of banana in Central America; black stem rust of wheat; southern corn leaf blight in the United States; Panama disease of banana in Asia, Australia, and Africa; and coffee rust in Central and South America. Such losses from plant diseases can have a significant economic impact, causing a reduction in income for crop producers and distributors and higher prices for consumers.

Loss of crops from plant diseases may also result in hunger and starvation, especially in less-developed countries where access to disease-control methods is limited and annual losses of 30 to 50 percent are not uncommon for major crops. In some years, losses are much greater, producing catastrophic results for those who depend on the crop for food. Major disease outbreaks among food crops have led to famines and mass migrations throughout history. The devastating outbreak of late blight of potato (caused by the water mold Phytophthora infestans) that began in Europe in 1845 brought about the Great Famine that caused starvation, death, and mass migration of the Irish. Of Ireland's population of more than eight million, approximately one million (about 12.5 percent) died of starvation or famine-related illness, and 1.5 million (almost 19 percent) emigrated, mostly to the United States, as refugees from the destructive blight. This water mold thus had a tremendous influence on economic, political, and cultural development in Europe and the United States. During World War I, late blight damage to the potato crop in Germany may have helped end the war.

Diseases—a normal part of nature

Plant diseases are a normal part of nature and one of many ecological factors that help keep the hundreds of thousands of living plants and animals in balance with one another. Plant cells contain special signaling pathways that enhance their defenses against insects, animals, and pathogens. One such example involves a plant hormone called jasmonate (jasmonic acid). In the absence of harmful stimuli, jasmonate binds to special proteins, called JAZ proteins, to regulate plant growth, pollen production, and other processes. In the presence of harmful stimuli, however, jasmonate switches its signaling pathways, shifting instead to directing processes involved in boosting plant defense. Genes that produce jasmonate and JAZ proteins represent potential targets for genetic engineering to produce plant varieties with increased resistance to disease.

Humans have carefully selected and cultivated plants for food, medicine, clothing, shelter, fibre, and beauty for thousands of years. Disease is just one of many hazards that must be considered when plants are taken out of their natural environment and grown in pure stands under what are often abnormal conditions.

Many valuable crop and ornamental plants are very susceptible to disease and would have difficulty surviving in nature without human intervention. Cultivated plants are often more susceptible to disease than are their wild relatives. This is because large numbers of the same species or variety, having a uniform genetic background, are grown close together, sometimes over many thousands of square kilometres. A pathogen may spread rapidly under these conditions.

Definitions of Plant Disease

In general, a plant becomes diseased when it is continuously disturbed by some causal agent that results in an abnormal physiological process that disrupts the plant's normal structure, growth, function, or other activities. This interference with one or more of a plant's essential physiological or biochemical systems elicits characteristic pathological conditions or symptoms.

Plant diseases can be broadly classified according to the nature of their primary causal agent, either infectious or noninfectious. Infectious plant diseases are caused by a pathogenic organism such as a fungus, bacterium, mycoplasma, virus, viroid, nematode, or parasitic flowering plant. An infectious agent is capable of reproducing within or on its host and spreading from one susceptible host to another. Noninfectious plant diseases are caused by unfavourable growing conditions, including extremes of temperature, disadvantageous relationships between moisture and oxygen, toxic substances in the soil or atmosphere, and an excess or deficiency of an essential mineral. Because noninfectious causal agents are not organisms capable of reproducing within a host, they are not transmissible.

In nature, plants may be affected by more than one disease-causing agent at a time. A plant that must contend with a nutrient deficiency or an imbalance between soil moisture and oxygen is often more susceptible to infection by a pathogen, and a plant infected by one pathogen is often prone to invasion by secondary pathogens. The combination of all disease-causing agents that affect a plant make up the disease complex. Knowledge of normal growth habits, varietal characteristics, and normal variability of plants within a species—as these relate to the conditions under which the plants are growing—is required for a disease to be recognized. The study of plant diseases is called plant pathology.

Disease Development and Transmission

Pathogenesis and Saprogenesis

Pathogenesis is the stage of disease in which the pathogen is in intimate association with living host tissue. Three fairly distinct stages are involved:

1. Inoculation: transfer of the pathogen to the infection court, or area in which invasion of the plant occurs (the infection court may be the unbroken plant surface, a variety of wounds, or natural openings—e.g. stomata [microscopic pores in leaf surfaces], hydathodes [stomata-like openings that secrete water], or lenticels [small openings in tree bark]).

2. Incubation: the period of time between the arrival of the pathogen in the infection court and the appearance of symptoms.

3. Infection: the appearance of disease symptoms accompanied by the establishment and spread of the pathogen.

One of the important characteristics of pathogenic organisms, in terms of their ability to infect, is virulence. Many different properties of a pathogen contribute to its ability to spread through and to destroy the tissue. Among these virulence factors are toxins that kill cells, enzymes that destroy cell walls, extracellular polysaccharides that block the passage of fluid through the plant system, and substances that interfere with normal cell growth. Not all pathogenic species are equal in virulence—that is, they do not produce the same amounts of the substances that contribute to the invasion and destruction of plant tissue. Also, not all virulence factors are operative in a particular disease. For example, toxins that kill cells are important in necrotic diseases, and enzymes that destroy cell walls play a significant role in soft rot diseases.

Many pathogens, especially among the bacteria and fungi, spend part of their life cycle as pathogens and the remainder as saprotrophs.

Saprogenesis is the part of the pathogen's life cycle when it is not in vital association with living host tissue and either continues to grow in dead host tissue or becomes dormant. During this stage, some fungi produce their sexual fruiting bodies; the apple scab (Venturia inaequalis), for example, produces perithecia, flask-shaped spore-producing structures, in fallen apple leaves. Other fungi produce compact resting bodies, such as the sclerotia formed by certain root- and stem-rotting fungi (Rhizoctonia solani and Sclerotinia sclerotiorum) or the ergot fungus (Claviceps purpurea). These resting bodies, which are resistant to extremes in temperature and moisture, enable the pathogen to survive for months or years in soil and plant debris in the absence of a living host.

Epiphytotics

When the number of individuals a disease affects increases dramatically, it is said to have become epidemic (meaning "on or among people"). A more precise term when speaking of plants, however, is epiphytotic ("on plants"); for animals, the corresponding term is epizootic. In contrast, endemic (enphytotic) diseases occur at relatively constant levels in the same area each year and generally cause little concern.

Epiphytotics affect a high percentage of the host plant population, sometimes across a wide area. They may be mild or destructive and local or regional in occurrence. Epiphytotics result from various combinations of factors, including the right combination of climatic conditions. An epiphytotic may occur when a pathogen is introduced into an area in which it had not previously existed. Examples of this condition include the downy mildews (Sclerospora species) and rusts (Puccinia species) of corn in Africa during the 1950s, the introduction of the coffee rust fungus into Brazil in the 1960s, and the entrance of chestnut blight (Endothia parasitica) into the United States shortly after 1900. Also, when new plant varieties are produced by plant breeders without regard for all enphytotic diseases that occur in the same area to some extent each year (but which are normally of minor importance), some of these varieties may prove very susceptible to previously unimportant pathogens. Examples of this situation include the development of oat varieties with Victoria parentage, which, although highly resistant to rusts (Puccinia graminis avenae and P. coronata avenae) and smuts (Ustilago avenae, U. kolleri), proved very susceptible to Helminthosporium blight (H. victoriae), formerly a minor disease of grasses. The destructiveness of this disease resulted in a major shift of oat varieties in the United States in the mid-1940s. Corn (maize) with male-sterile cytoplasm (i.e., plants with tassels that do not extrude anthers or pollen), grown on 60 million acres (24 million hectares) in the United States, was attacked in 1970 by a virulent new race of the southern corn leaf blight fungus (Helminthosporium maydis race T), resulting in a loss of about 700 million bushels of corn. More recently the new Helminthosporium race was widely disseminated and was reported from most continents. Finally, epiphytotics may occur when host plants are cultivated in large acreages where previously little or no land was devoted to that crop.

Epiphytotics may occur in cycles. When a plant disease first appears in a new area, it may grow rapidly to epiphytotic proportions. In time, the disease wanes, and, unless the host species has been completely wiped out, the disease subsides to a low level of incidence and becomes enphytotic. This balance may change dramatically by conditions that favour a renewed epiphytotic. Among such conditions are weather (primarily temperature and moisture), which may be very favourable for multiplication, spread, and infection by the pathogen; introduction of a new and more susceptible host; development of a very aggressive race of the pathogen; and changes in cultural practices that create a more favourable environment for the pathogen.

Environmental Factors affecting Disease Development

Important environmental factors that may affect development of plant diseases and determine whether they become epiphytotic include temperature, relative humidity, soil moisture, soil pH, soil type, and soil fertility.

Temperature

Each pathogen has an optimum temperature for growth. In addition, different growth stages of fungi, such as the production of spores (reproductive units), their germination, and the growth of the mycelium (the filamentous main fungus body), may have slightly different optimum temperatures. Storage temperatures for certain fruits, vegetables, and nursery stock are manipulated to control fungi and bacteria that cause storage decay, provided the temperature does not change the quality of the products. Little, except limited frost protection, can be done to control air temperature in fields, but greenhouse temperatures can be regulated to check disease development.

Knowledge of optimum temperatures, usually combined with optimum moisture conditions, permits forecasting, with a high degree of accuracy, the development of such diseases as blue mold of tobacco (Peronospora tabacina), downy mildews of vine crops (Pseudoperonospora cubensis) and lima beans (Phytophthora phaseoli), late blight of potato and tomato (Phytophthora infestans), leaf spot of sugar beets (Cercospora beticola), and leaf rust of wheat (Puccinia recondita tritici). Effects of temperature may mask symptoms of certain viral and mycoplasmal diseases, however, making them more difficult to detect.

Relative Humidity

Relative humidity is very critical in fungal spore germination and the development of storage rots. Rhizopus soft rot of sweet potato (Rhizopus stolonifer) is an example of a storage disease that does not develop if relative humidity is maintained at 85 to 90 percent, even if the storage temperature is optimum for growth of the pathogen. Under these conditions, the sweet potato root produces suberized (corky) tissues that wall off the Rhizopus fungus.

High humidity favours development of the great majority of leaf and fruit diseases caused by fungi, water molds, and bacteria. Moisture is generally needed for spore germination, the multiplication and penetration of bacteria, and the initiation of infection. Germination of powdery mildew spores occurs best at 90 to 95 percent relative humidity. Diseases in greenhouse crops—such as leaf mold of tomato (Cladosporium fulvum) and decay of flowers, leaves, stems, and seedlings of flowering plants, caused by Botrytis species—are controlled by lowering air humidity or by avoiding spraying plants with water.

Soil Moisture

High or low soil moisture may be a limiting factor in the development of certain root rot diseases. High soil-moisture levels favour development of destructive water mold fungi, such as species of Aphanomyces, Pythium, and Phytophthora. Excessive watering of houseplants is a common problem. Overwatering, by decreasing oxygen and raising carbon dioxide levels in the soil, makes roots more susceptible to root-rotting organisms.

Diseases such as take-all of cereals (Ophiobolus graminis); charcoal rot of corn, sorghum, and soybean (Macrophomina phaseoli); common scab of potato (Streptomyces scabies); and onion white rot (Sclerotium cepivorum) are most severe under low soil-moisture levels.

potato scab :Common scab of potato

Soil pH

Soil pH, a measure of acidity or alkalinity, markedly influences a few diseases, such as common scab of potato and clubroot of crucifers (Plasmodiophora brassicae). Growth of the potato scab organism is suppressed at a pH of 5.2 or slightly below (pH 7 is neutral; numbers below 7 indicate acidity, and those above 7 indicate alkalinity). Scab is not normally a problem when the natural soil pH is about 5.2. Some farmers add sulfur to their potato soil to keep the pH about 5.0. Clubroot of crucifers (members of the mustard family, including cabbage, cauliflower, and turnips), on the other hand, can usually be controlled by thoroughly mixing lime into the soil until the pH becomes 7.2 or higher.

Soil type

Certain pathogens are favoured by loam soils and others by clay soils. Phymatotrichum root rot attacks cotton and some 2,000 other plants in the southwestern United States. This fungus is serious only in black alkaline soils—pH 7.3 or above—that are low in organic matter. Fusarium wilt disease, which attacks a wide range of cultivated plants, causes more damage in lighter and higher (topographically) soils. Nematodes are also most damaging in lighter soils that warm up quickly.

Soil Fertility

Greenhouse and field experiments have shown that raising or lowering the levels of certain nutrient elements required by plants frequently influences the development of some infectious diseases—for example, fire blight of apple and pear, stalk rots of corn and sorghum, Botrytis blights, Septoria diseases, powdery mildew of wheat, and northern leaf blight of corn. These diseases and many others are more destructive after application of excessive amounts of nitrogen fertilizer. This condition can often be counteracted by adding adequate amounts of potash, a fertilizer containing potassium.

Requirements for Disease Development

Infectious disease cannot develop if any one of the following three basic conditions is lacking: (1) the proper environment, the most important environmental factors being the amount and frequency of rains or heavy dews, the relative humidity, and the air and soil temperatures, (2) the presence of a virulent pathogen, and (3) a susceptible host. Effective disease-control measures are aimed at breaking this environment-pathogen-host triangle. Loss resulting from disease is reduced, for example, if the host can be made more resistant or immune through such techniques as plant breeding or genetic engineering. In addition, the environment can be made less favourable for invasion by the pathogen and more favourable for the growth of the host plant. Finally, the pathogen can be killed or prevented from reaching the host. These basic methods of control can be divided into a number of cultural, chemical, and biological practices to help control the disease.

Classification of Plant Diseases by Causal Agent

Plant diseases are often classified by their physiological effects or symptoms. Many diseases, however, produce practically identical symptoms and signs but are caused by very different microorganisms or agents, thus requiring completely different control methods. Classification according to symptoms is also inadequate because a causal agent may induce several different symptoms, even on the same plant organ, which often intergrade. Classification may be according to the species of plant affected. Host indexes (lists of diseases known to occur on certain hosts in regions, countries, or continents) are valuable in diagnosis. When an apparently new disease is found on a known host, a check into the index for the specific host often leads to identification of the causal agent. It is also possible to classify diseases according to the essential process or function that is adversely affected. The best and most widely used classification of plant diseases is based on the causal agent, such as a noninfectious agent or an infectious agent (i.e., a virus, viroid, mycoplasma, bacterium, fungus, nematode, or parasitic flowering plant).

Noninfectious Disease-causing Agents

Noninfectious diseases, which sometimes arise very suddenly, are caused by the excess, deficiency, nonavailability, or improper balance of light, air circulation, relative humidity, water, or essential soil elements; unfavourable soil moisture-oxygen relations; extremes in soil acidity or alkalinity; high or low temperatures; pesticide injury; other poisonous chemicals in air or soil; changes in soil grade; girdling of roots; mechanical and electrical agents; and soil compaction. In addition, unfavourable preharvest and storage conditions for fruits, vegetables, and nursery stock often result in losses. The effects of noninfectious diseases can be seen on a variety of plant species growing in a given locality or environment. Many diseases and injuries caused by noninfectious agents result in heavy loss but are difficult to check or eliminate because they frequently reflect ecological factors beyond human control. Symptoms may appear several weeks or months after an environmental disturbance.

Injuries incurred from accidents, poisons, or adverse environmental disturbances often result in damaged tissues that weaken a plant, enabling bacteria, fungi, or viruses to enter and add further damage. The cause may be obvious (lightning or hail), but often it is obscure. Symptoms alone are often unreliable in identifying the causal factor. A thorough examination of recent weather patterns, the condition of surrounding plants, cultural treatments or disturbances, and soil and water tests can help reveal the nature of the disease.

Adverse Environment

High temperatures may scald corn, cotton, and bean leaves and may induce formation of cankers at the soil surface of tender flax, cotton, and peanut plants. Frost injury is relatively common, but temperatures just above freezing also may cause damage, such as net necrosis (localized tissue death) in potato tubers and "silvering" of corn leaves. Isolated, thin-barked trees growing in northern climates and subjected to frequent thawing by day and freezing by night may develop dead bark cankers or vertical frost cracks on the south or southwest sides of the trunk. Alternate freezing and thawing, heaving, low air moisture, and smothering under an ice-sheet cover are damaging to alfalfa, clovers, strawberries, and grass on golf greens. Legume crowns commonly split under these conditions and are invaded by decay-forming fungi.

The drought and dry winds that often accompany high temperatures cause stunting, wilting, blasting, marginal scorching of leaves, and dieback of shoots. Leaf scorch is common on trees in exposed locations following hot, dry, windy weather when water is lost from leaves faster than it is absorbed by roots. Leaf scorch and sudden flower drop are common indoor plant problems because the humidity in a home, an apartment, or an office is usually below 30 percent. Similar symptoms are caused by a change in soil grade, an altered water-table level, a compacted and shallow soil, paved surface over tree roots, temporary flooding or a waterlogged (oxygen-deficient) soil, girdling tree roots, salt spray near the ocean, and an injured or diseased root system. Injured plants are often very susceptible to air and soil pathogens and secondary invaders.

Blossom-end rot of tomato and pepper is prevalent when soil moisture and temperature levels fluctuate widely and calcium is low.

Poor aeration may cause blackheart in stored potatoes. Accumulation of certain gases from the respiration of apples in storage may produce apple scald and other disorders.

All plants require certain mineral elements to develop and mature in a healthy state. Macronutrients such as nitrogen, potassium, phosphorus, sulfur, calcium, and magnesium are required in substantial quantities, while micronutrients or trace elements such as boron, iron, manganese, copper, zinc, and molybdenum are needed in much smaller quantities. When the supply of any essential nutrient falls below the level required by the plant, a deficiency occurs, leading to symptoms that include stunting of plants; scorching or malformation of leaves; abnormal coloration; premature leaf, bud, and flower drop; delayed maturity or failure of flower and fruit buds to develop; and dieback of shoots.

Symptoms of nutrient deficiencies vary depending on the nutrients involved, the stage of plant growth, soil moisture, and other factors; they often resemble symptoms caused by infectious agents such as bacteria or viruses.

The availability of water may affect nutrient uptake by the plant. Blossom-end rot of tomato, a disease associated with a deficiency of calcium, may occur if the water supply is irregular, even if an adequate amount of calcium is in the soil. This discontinuity in availability of water will inhibit uptake of the calcium in a quantity sufficient to nourish a fast-growing tomato plant. Necrosis at the blossom end of the fruit results. This situation generally disappears when water conditions improve.

Excess minerals can damage plants either directly, causing stunting, deformities, or dieback, or indirectly by interfering with the absorption and use of other nutrients, resulting in subsequent deficiency symptoms. A superabundance of nitrogen, for example, may cause deficiency symptoms of potassium, zinc, or other nutrient elements; a lack of or delay in flower and fruit development;

and a predisposition to winter injury. If potassium is high, calcium and magnesium deficiencies may occur.

The pH of a soil has a dramatic impact on nutrient availability to plants. Most plants will grow in a soil with a pH between 4.0 and 8.0. In acidic soils some nutrients are far more available and may reach concentrations that are toxic or that inhibit absorption of other nutrients, while other minerals become chemically bound and unavailable to plants. A similar situation exists in alkaline soils, although different minerals are affected. Oats planted in alkaline soils that actually contain a sufficient amount of manganese may develop the manganese-deficiency disease gray speck. This occurs because an elevated soil pH causes manganese to react with oxygen to produce manganese dioxide, a form of the nutrient that is insoluble to plants.

An excess of water-soluble salts is a common problem with houseplants. Salt concentrations may build up as a whitish crust on soil and container surfaces of potted plants following normal evaporation of water over a period of time. Symptoms include leaf scorching, bronzing, yellowing and stunting, and wilting, plus root and shoot dieback. Damage from soluble salts is also common in arid regions and in regions where ice-control chemicals are applied heavily.

Several nonparasitic diseases (e.g. oat blast, weakneck of sorghum, straighthead of rice, and crazy-top of cotton) are caused by combinations of environmental factors—e.g. high temperatures, moisture stress or poor irrigation practices, imbalance of mineral nutrients, and reduced light.

Environmental disturbances alter the normal physiology of the plant, activity of pathogens, and host-pathogen interactions.

Toxic Chemicals

Many complex chemicals are routinely applied to plants to prevent attack by insects, mites, and pathogens; to kill weeds; or to control growth. Serious damage may result when fertilizers, herbicides, fumigants, growth regulators, antidesiccants, insecticides, miticides, fungicides, nematicides, and surfactants (substances with enhanced wetting, dispersing, or cleansing properties, such as detergents) are applied at excessive rates or under hot, cold, or slow-drying conditions.

Some pollutants are the direct products of industry and fuel combustion, while others are the result of photochemical reactions between products of combustion and naturally occurring atmospheric compounds. The major pollutants toxic to plants are sulfur dioxide, fluorine, ozone, and peroxyacetyl nitrate.

Sulfur dioxide results primarily from the burning of large amounts of soft coal and high-sulfur oil. It is toxic to a wide range of plants at concentrations as low as 0.25 part per million (ppm) of air (i.e., on a volume basis, one part per million represents one volume of pure gaseous toxic substance mixed in one million volumes of air) for 8 to 24 hours. Gaseous and particulate fluorides are more toxic to sensitive plants than is sulfur dioxide because they are accumulated by leaves. They are also toxic to animals that feed on such foliage. Fluorine injury is common near metal-ore smelters, refineries, and industries making fertilizers, ceramics, aluminum, glass, and bricks.

Ozone and peroxyacetyl nitrate injury (also called oxidant injury) are more prevalent in and near cities with heavy traffic problems. Exhaust gases from internal combustion engines contain large amounts of hydrocarbons (substances that principally contain carbon and hydrogen molecules—gasoline, for example). Smaller amounts of unconsumed hydrocarbons are formed by combustion of fossil fuels (e.g. coal, oil, natural gas) and refuse burning. Ozone, peroxyacetyl nitrate, and other oxidizing chemicals (smog) are formed when sunlight reacts with nitrogen oxides and hydrocarbons. This pollutant complex is damaging to susceptible plants many kilometres from its source. Ozone and peroxyacetyl nitrate are capable of causing injury if present at levels of 0.01 to 0.05 part per million for several hours.

Physical Injury

Lightning, hail, high winds, ice and snow loads, machinery, insect and animal feeding, and various cultural practices may seriously injure plants or plant products. With the exception of lightning, which may cause death of trees and succulent crop plants in limited areas, such injury does not usually kill plants. Wounds are created, however, through which pathogens may enter.

Infectious Disease-causing Agents

Plants are subject to infection by thousands of species from very diverse groups of organisms. Most are microscopic, but a few are macroscopic. The infectious agents, as previously mentioned, are called pathogens and can be grouped as follows: viruses and viroids, bacteria (including mycoplasmas and spiroplasmas, collectively referred to as mycoplasma-like organisms [MLOs]), fungi, nematodes, and parasitic seed plants.

Diseases Caused by Viruses and Viroids

Viruses and viroids are the smallest of the infectious agents. The structurally mature infectious particle is called a virion. Virions range in size from approximately 20 nanometres (0.0000008 inch) to 250–400 nanometres and are of various shapes. Viroids differ from viruses in that they have no structural proteins, such as those that form the protein coat (capsid) of the virus.

Both viruses and viroids are obligate parasites—i.e., they are able to multiply or replicate only within a living cell of a particular host. A single plant species may be susceptible to infection by several different viruses or viroids. Major disease of important food crops such as potato, tomato, wheat, oats, rice, corn, peach, orange, sugar beet, sugarcane, and palm result from viral infection. Diseases are generally most serious in plants that are propagated vegetatively, or asexually—i.e., grown from cuttings, cut divisions, sprouts, and other plant material—rather than grown from seeds (sexually propagated).

Symptoms

The symptoms of viral and viroid plant diseases fall into four groups: (1) change in colour—yellowing, green and yellow mottling, and vein clearing; (2) malformations—distortion of leaves and flowers, rosetting, proliferation and witches'-brooms (abnormal proliferation of shoots), and little or no leaf development between the veins; (3) necrosis—leaf spots, ring spots, streaks, wilting or drooping, and internal death, especially of phloem (food-conducting) tissue; and (4) stunting or dwarfing of leaves, stems, or entire plants. Rarely they may kill the host in a short time (e.g. spotted wilt and curly top of tomato). More commonly they cause reduced yield and lower quality of product.

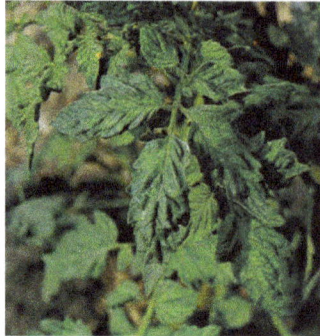

Tomato leaves puckered and blistered by the tobacco mosaic virus.

In many cases, virus-infected plants are more susceptible to root rots, stem or stalk rots, seedling blights, and possibly other types of diseases.

Some plants may carry one or more viruses and show no symptoms; thus, they are latent carriers and a source of infection for other plants. Symptoms of certain virus-infected plants, such as geraniums, may be masked at high temperatures. Virus symptoms reappear when the weather cools.

For convenience, viral/viroid diseases are often grouped together generally by symptoms, regardless of true viral/viroid relationships. Viruses also can be grouped into strains, each differing greatly in virulence and other properties. For example, two virus strains, chemically distinct, may produce indistinguishable symptoms in one orchid plant but strikingly different symptoms in another. Diseases caused by unrelated viruses may resemble one another more closely than diseases caused by strains of the same virus. Certain variegated plants, such as Abutilon and Rembrandt tulips, owe their horticultural uniqueness and desirability to being inherently virus-infected.

Transmission

With the exception of tobacco mosaic virus, relatively few viruses or viroids are spread extensively in the field by contact between diseased and healthy leaves.

All viruses that spread within their host tissues (systemically) can be transmitted by grafting branches or buds from diseased plants on healthy plants. Natural grafting and transmission are possible by root grafts and with parasitic dodder (Cuscuta species). Vegetative propagation often spreads plant viruses. Fifty to 60 viruses are transmitted in seed, and a few seed-borne viruses, such as sour-cherry yellows, are carried in pollen and transmitted by insects.

Most disease-causing viruses are carried and transmitted naturally by insects and mites, which are called vectors of the virus. The principal virus-carrying insects are about 200 species of aphids, which transmit mostly mosaic viruses, and more than 100 species of leafhoppers, which carry yellows-type viruses. Whiteflies, thrips, mealybugs, plant hoppers, grasshoppers, scales, and a few beetles also serve as vectors for certain viruses. Some viruses may persist for weeks or months and even duplicate themselves in their insect vectors; others are carried for less than an hour. Slugs, snails, birds, rabbits, and dogs also transmit a few viruses, but this is not common.

A small number of plant viruses are soilborne. Viruses causing grape fanleaf, tobacco rattle, and tobacco and tomato ring spots, as well as several strawberry viruses, are spread by nematodes feeding externally (i.e., ectoparasitic) on plant roots. A few soilborne viruses may be spread by the swimming spores of primitive, soil-inhabiting pathogenic fungi, such as those causing big vein of lettuce, soilborne wheat mosaic, and tobacco necrosis.

Viruses often overwinter in biennial and perennial crops and weeds (plants that overwinter by means of roots and produce seed in their second year or during several years, respectively), in plant debris, and in insect vectors. Plants, once infected, normally remain so for life.

Control

After a plant is infected with a virus/viroid, little can be done to restore its health. Control is accomplished by several methods, such as growing resistant species and varieties of plants or obtaining virus-free seed, cuttings, or plants as a result of indexing and certification programs. Indexing is a procedure to determine the presence or absence of viruses not readily transmitted mechanically. Material from a "test" plant is grafted to an "indicator" plant that develops characteristic symptoms if affected by the viral disease in question. In addition, more drastic measures are sometimes followed, including destroying (roguing) infected crop and weed host plants and enforcing state and national quarantines or embargoes. Further control measures include controlling insect vectors by spraying plants with contact insecticides or fumigating soil to kill insects, nematodes, and other possible vectors. Growing valuable plants under fine cheesecloth or wire screening that excludes insect vectors also is done. Separation of new from virus-infected plantings of the same or closely related species is sometimes effective, and the simple practice of not propagating from plants suspected or known to harbour a virus also reduces loss.

Infected peach, apple, and rose budwood stock and carnations have been grown for weeks or months at temperatures about 37 to 38 °C (99 to 100 °F) to free new growth from viruses. Soaking some woody plant parts or virus-infected sugarcane shoots in hot water at about 50 °C (120 °F) for short periods also is effective. Both dry and wet heat treatments are based on the sensitivity of certain viruses to high temperatures. Rapidly growing dahlia and chrysanthemum sprouts outgrow viruses so that stem tips can be used to propagate healthy plants. With certain carnations, chrysanthemums, and potatoes, a few cells from the growing tip have been grown under sterile conditions in tissue culture; from these, whole plants have been developed free from viruses.

Examples of virus and viroid diseases are characterized in the table.

Some viral and viroid diseases of plants				
disease	causative agent	hosts	symptoms and signs	additional features
tobacco mosaic	tobacco mosaic virus (TMV)	tobacco, tomato, and hundreds of other vegetables and weeds	mottled appearance of leaves (mosaic pattern); dwarfing	virus remains viable for years in soil and tobacco; the disease occurs world-wide; significant economic losses can occur

cucumber mosaic	cucumber mosaic virus (CMV)	cucumber, bean, tobacco, and other plants (wide range of hosts)	similar to those of TMV infections	worldwide occurrence; very broad range of hosts
barley yellow dwarf	barley yellow dwarf virus (BYDV)	barley, oats, rye, wheat; also pasture grasses and weeds	yellowing and dwarfing of leaves; stunting of plants	one of the most important diseases of small grains
tomato spotted wilt	tomato spotted wilt virus (TSWV)	tomato, pepper, pineapple, peanut, and many other plants	leaves show concentric, necrotic rings; necrotic region yellow, then turning red-brown	very wide host range; infects hundreds of different plants
prunus necrotic ring spot	prunus necrotic ring spot virus (PNRV)	stone fruits—e.g. cherry, almond, peach, apricot, plum, and others	delayed foliation; leaves on infected branches show light green spots and dark rings, then become necrotic and fall off	very widespread disease of stone fruits; affects almost all trees in fruit-producing regions
potato spindle tuber	potato spindle tuber viroid (PSTV)	potato and tomato	stunted growth; tubers are spindle-shaped and smaller than healthy tubers	the first identified viroid infection in plants; can cause major reduction in crop yield
citrus exocortis	citrus exocortis viroid (CEV)	orange, lemon, lime, and other citrus plants	infected trees show vertical splits in bark, thin strips of partially loosened bark, and a cracked, scaly appearance	worldwide distribution; causes reduction of crop yield

Plant Pathogens

A plant pathogen is a broad term that refers to any of the organisms, such as fungi, bacteria, protists, nematodes, and viruses that cause plant diseases. Plant pathogens are of interest for a number of reasons, ranging from concerns about fragile ecosystems to the desire to protect the food supply. Plant pathogens that cause plant diseases reduce a grower's ability to produce crops and can infect almost all types of plants.

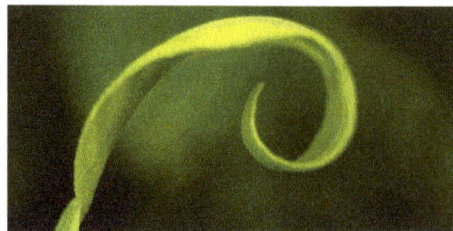

Plant Pathogen Symptoms

Plants are attacked by different groups of pathogens individually or sometimes by more than one pathogen-producing complex and more severe disease. The type of external symptoms can, in most cases, indicate the nature of the pathogen responsible for the disease. Plant pathogens can attack in a number of different ways. Some colonize the tissue in the plant, others settle on the surface of the plant, and others may go for specific areas such as the roots, stems, and leaves. Pathogens commonly cause problems like tissue death, browning, a decrease in fruiting, problems with setting flowers, and so forth. In extreme cases, they can kill the host plant.

Here are a few examples of common signs and symptoms of fungal, bacterial and viral plant diseases:

Fungal disease symptoms	Bacterial disease symptoms	Viral disease symptoms
Birds-eye spot on berries (anthracnose)	Leaf spot with yellow halo Fruit spot	Mosaic leafpattern
Damping off of seedlings (phytophthora)	Canker	Crinkled leaves
Leaf spot (septoria brown spot)	Crown gall	Yellowed leaves
Chlorosis (yellowing of leaves)	Sheperd's crook stem ends on woody plants	Plant stunting

Bioassay is a useful procedure for detection and identification of viruses and uses indicator plants which react to infection by showing characteristic symptoms. Over the last few decades, laboratory based virus test methods, such as Enzyme-Linked-Immuno-Sorbent Assay (ELISA) and Polymerase Chain Reaction (PCR), have been developed and are now being used routinely in many laboratories. Automation of these procedures enables a high throughput of samples and provides rapid results.

Plant Disease Triangle: Plant pathologists have identified three factors that are needed for a plant disease to develop:

i.   A susceptible host Some pathogens have a narrow host plant range, meaning they can infect just a few host species. For instance, the primary host crops of Late blight (Phytophthora infestans) are tomato and potato. By contrast, pathogens with a wide host plant range can infect many different host species. There are almost 200 plant species that can be infected by Bacterial wilt (Ralstonia solanacearum).

ii. A disease-causing organism (pathogen). Plant pathogens include fungi, bacteria, viruses, and nematodes.

iii. A favorable environment for the pathogen. Pathogens usually require specific humidity and temperature conditions for pathogen infection and disease symptoms to manifest. For instance, Late Blight disease symptoms are most likely to occur when the weather is cool and wet.

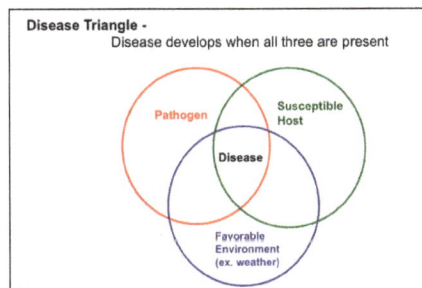

Disease Triangle

Disease Diagnosis

The three disease triangle factors are important for diagnosing the cause of disease symptoms. Pathologists consider the weather, environmental conditions and the host species to diagnosis what pathogen is causing disease symptoms. Pathologists also consider other factors that could favor and help diagnose a disease, such as i. the field history, particularly what crops and pathogens were present in the past, ii. current crop management practices, iii. when disease symptoms were visible, and on what other species. To assist farmers and others with disease diagnosis, many land-grant universities in the US have crop and animal diagnostic disease clinics where one can submit diseased tissue samples with detailed information that can aid in the diagnosis, such as the host species, environmental conditions, the site history, and management.

Pathogen Management

Although disease control practices could be categorized into the pest control approaches that were discussed earlier for managing insects and weeds (genetic, cultural, chemical, etc.), plant pathologists typically describe pathogen control tactics with more specific language. For instance, Exclusion tactics involve rejecting infected transplants from being introduced to a farm.

Prevention or Avoidance of pathogen introduction and spread tactics include:

- Crop rotation, particularly for pathogens with narrow host ranges.

- Sanitizing equipment for planting, trellising, pruning, and harvesting.

- Managing for healthy vigorous crops with optimal soil and water nutrient management.

- Managing to avoid environmental conditions that promote pathogens, such as avoiding very humid conditions due to over-watering, or promoting drying of plant surfaces with wide row spacing to facilitate air flow, or using drip-tape irrigation that waters plants at or below the soil surface versus over the canopy.

- Using physical barriers such as row covers, mulch to reduce water splashing; and high tunnel/ hoop houses or greenhouses to prevent the introduction of rain and wind-borne pathogens.

Multiple pest control

Multiple pest control practices on this organic farm help prevent pathogen infection and spread, while also helping to control insect pests and weeds. Wide row spacing allows for interrow weed cultivation as well as air flow to reduce crop canopy humidity; crop rotation and intercropping plants from different plant families interrupts pathogen and insect spread; and straw mulch that prevents soil-borne pathogens from splashing onto plants also suppresses weeds. Plastic mulch also raises soil temperatures, promoting crop growth and helps to suppress weeds.

Drip tape irrigation under plastic mulch avoids splashing soil-borne pathogens onto crops and is also a more efficient use of irrigation water.

High tunnels, plastic mulch, and sanitized stakes all avoid introducing pathogens to these tomato plants, while also promoting crop growth.

Genetic resistanceto pathogens is a very valuable and important pathogen control tool. Many plant breeding programs select for genetic resistance to pathogens. When available, pathogen resistance traits are included in most crop variety descriptions to help growers select appropriate crop varieties for their farm.

This tomato variety trial included tomato varieties that were resistant or susceptible to late blight.

If disease symptoms develop, infected plants may be Eradicated or destroyed. And materials that may have been contaminated with pathogens, such as the soil and planting containers, can be heated to very high temperatures with pasteurization equipment or through solarization. For instance, soil may be solarized by placing black plastic over the crop bed (planting zone) during the warm season to increase the soil temperature and destroy pathogens prior to planting the crop.

Therapy or Fungicides(chemical control) may be applied to infected plants to terminate pathogens. Particularly when plant pathogen symptoms are identified early and favorable weather conditions for the pathogen are projected to continue, fungicides can prevent disease spread and significant economic losses. In some high-value crop systems, the soil may be fumigated prior to planting crops.

Similarly, in agricultural livestock systems, animals with disease symptoms can be treated with antibiotics. And in some livestock production systems, antibiotics and vaccinations are administered to animals to prevent diseases and pathogen infection.

Types of Penetration of Post-Harvest Pathogens

Infield Penetration

Fungi and bacteria responsible for in-storage decay often originate in the field or orchard. Late blight of potatoes caused by Phytophthora infestans is an example of decay originating in tuber infection in the field. The infection is caused by the zoospores found in soil or that fall onto the tubers from infected foliage during harvest.

Following germination, the zoospores penetrate into the tubers through eyes, lenticels, growth cracks, wounds, or via the point of attachment to the plant. Tubers that were infected a few days prior to harvest or during the harvest itself are brought to storage carrying the disease in its early developmental stages, with no visible symptoms of decay.

In a warm and damp climate, the blight fungus can also attack the tomato fruit at its various ripening stages in the field. The attack usually takes place at the edges of the fruit stalk or directly through the skin. In course of epidemics the entire crop may rot in the field.

However, when the disease is less severe, fruits with no visible symptoms or with slight blemishes might be picked and the fungus will continue to develop during storage.

The brown rot, which develops in citrus fruits during storage, originates in preliminary infections initiated in the orchards by Phytophthora citrophtliora and other Phytophthora species. During the rainy season, the fungal zoospores descend on the lower fruits of the tree and penetrate them directly.

The fungus can develop in the orchard, but zoospore infection occurs a few days prior to picking and the external symptoms are not visible, the fruits will later rot and might comprise a serious problem during storage or transportation.

The principle concern is that the harvest ends fungus development within the host while it is still on the parent plant and causing the decay after harvest or during storage. In many vegetables such as carrot, cucumber, lettuce, celery, cabbage, cauliflower and others, the soft watery rot caused by Sclerotinia sclerotiorum is common.

This fungus, common in the soil might attack the vegetable when still in the field or during harvest and continue to develop in its tissues throughout storage or marketing.

Latent Infections

Pathogen penetration takes place in the field and will get to storeroom within the host tissue without eliciting any symptoms of decay known as early or quiescent stages of infection. The fungi that penetrate into the host in fields may cause latent or quiescent infections. These pathogens arrive at fruits or vegetables that are still on parent plant.

However, during one of the phases, between their reaching the host and development of progressive disease, the growth is arrested until after the harvest, when physiological and biochemical changes occurring within the host will enable their renewed growth. Verhoeff has described such arrested infections as "latent infections" in which the pathogen growth is temporarily inhibited.

The latent state of the pathogen, whether involving spores that have landed on the host surface have commenced germination, or primary hyphal development within the host tissues, is linked to a dynamic balance among the host, the pathogen and the environment.

The physiological and biochemical changes occurring in the host tissues after and during storage might affect the pathogen and host interrelationship lead to the activation of the latent pathogen. The brown rot, caused by Monilinia fructicola in stone fruits and the decay caused by Nectria galligena in apples originate in the infections of young fruit in orchard.

The Nectria fungus forms a preliminary mycelium within the young tissues, which is the dormant stage of disease, while the interrupted stage in the development of the brown rot caused by Monilinia occur during spore penetration through the stomata or even at the pre-germination stage.

Another example of quiescent infection is anthracnose caused by Colletotrichum gloeosporioides in many tropical and subtropical fruits such as mango,guava, papaya and various citrus fruits. The fungal conidia are found in large quantities on the surface of fruits during their development on the tree.

In presence of free water upon the fruit, the spores germinate and form an appressorium at the tip of the germ tube. The appressoria function as resting spores on the fruit, since they resist environmental conditions much better than the conidia and remain vital for long periods while embedded in the natural wax of the fruit or bound to its surface.

Contact Infections

Fruits or vegetables which are free of a pathogen invasion by any means of penetration might be infected through contact with infected produce during storage. Contact infection is a significant factor in spreading white watery rot (Sclerotinia spp.) and bacterial soft rot (Erwinia spp.) in lettuce, carrot, cabbage, celery or squash during storage.

The development of Botrytis infection in stored strawberries, which turns into "mummies" covered with a gray layer of spore-bearing mycelium, causes contact infection and makes fruits susceptible to the disease. Similarly, one strawberry or tomato or a single grape infected by Rhizopus can spread the infection within the container.

In fact, contact infection by Botrytis or Rhizopus is typical of many fruits and vegetables and account for major losses caused by these pathogens during long-term storage. In citrus fruits, contact infection of green and blue mould caused by Penicillium digitatum and P. italicum is very common; it often occurs during shipment and under certain conditions disqualifies the entire shipment.

Penetration through Natural Openings

Some pathogenic fungi and bacteria that cannot normally penetrate the directly, without the presence of wound in its surface, can penetrate through natural openings such as stomata and lenticels.

The penetration of germ tubes of Colletotrichum gloeosporioides spores into young papaya fruits and the penetration of Monilinia fructicola spore germ tubes into young stone fruits can take place through stomata while the fruit is still in the orchard.

Penetration through lenticels has also been described for Alternaria alternata spores in mango and Gloeosporium perennans in apples. The lenticels in potato tubers are prone to develop bacterial soft rot during storage. Most of lenticels are infested with the cells of Erwinia cartovora during harvest.

The bacteria remain inactive in the lenticels until the favourable conditions such as mechanical pressure, existence of free water or low oxygen pressure within the tuber enhancing the tuber sensitivity to decay.

Penetration through Wounds

In contrast to pathogens that attack the fruit and vegetable in the field, most of the storage pathogens are incapable of penetrating directly through cuticle or epidermis of the host, but require a wound or an injury to facilitate their penetration. Therefore, the fungi and bacteria that develop during storage are often called wound pathogens.

Growth cracks present on harvested fruits and vegetables are natural avenues of infections. Careless separation of the fruit or vegetable from the parent plant might result in an injury liable to be attacked by the pathogen. The extent of injury caused by mechanical harvesting is far greater than that caused by a manual operation.

The harvesting is accompanied by mechanical injuries enables the weak pathogens to penetrate. It has turned out that the depth of the injury, combined with the humid conditions during storage, determine the fate and extent of the infection.

Each scratch, incision or other mechanical injuries inflicted on the fruit or vegetable during handling, harvesting, transporting, sorting, packing and storing might present adequate penetration points for the storage pathogens. A possible penetration point is the stem end separation area where damage often occurs during fruit picking and in this regard, the separation area is no different from any other injury.

It regularly happens that simultaneously with the injury, a large amount of fungal spores and bacterial cells arrive at the injured area, some of which use the injury site to penetrate and infect the host. Penetration through wounds is a characteristic of Penicillium digitatum and Penicillium italicum pathogen conidia in citrus fruits.

Wound infection is also characteristic of Rhizopus stolonifer that causes watery soft rot in many species of fruits and vegetables; of Alternaria alternata that causes a dark, rather dry decay in a large number of fruits and vegetables stored for a prolonged period; of Geotriclium candidum which causes the sour rot in citrus, melons and tomato fruits; and of various Aspergillus, Cladosporium and Trichothecium species, and other storage fungi of various fruits and vegetables.

Penetration Owing to Result of Senescence

Tissue senescence during continuing storage reduces disease resistance. Commonly at the end of storage period, the sensitivity of melons to blue green mould caused by various species of Penicillium and pink mould caused by Trichothecium roseum is increased. A senescencing onion that has commenced sprouting often harbours base decay, caused by various species of Fusarium.

Generally, the rate of decay during storage increases with the duration of storage as tissue senescence progresses. Increasing the tissue sensitivity to diseases during storage also contributes to contact infection of a healthy product by an infected one covered with spore-bearing mycelium.

Penetration after Physical Damage

Injuries caused by low temperature, heat, oxygen shortage or any other environmental stress increases the fruit or vegetable sensitivity to storage fungi. The physiological damage can be externally expressed through tissue browning and splitting, thus forming site vulnerable to invasion of wound pathogens.

Extreme environmental conditions enhance sensitivity to an attack without any visible external signs of damage. Development of sun scald lesions in apples leads to Alternaria alternata and Stemphylium hotryosum attack while Alternaria may also be associated with other physiological disorders on apples such as bitter pit or soft scald.

Plant-Pathogen Interaction

During the contact between plant and pathogenic microorganism, a particular chain of events is produced in the plant organism. The interaction between plant and pathogen may develop by two ways given below.

1. The plant is provided by a receptor that interacts with bacterial protein. As a result, quick protective reaction is being developed. In such a situation, the bacteria is called avirulent for a given plant genotype.

2. The proteins of the pathogenic organism are virulent for the given plant genotype. The plant is affected by the pathogen, whereas protective mechanisms are being activated more slowly.

In both cases, with the start of pathogenesis gene transcription, the cell walls strengthen. Then in the place of pathogen penetration, the active forms of oxygen are formed, causing the death of infected cells.

At the present moment, the best studied are the molecular mechanisms providing hypersensitive response (HR). In this case, the plant receptor interacts with the pathogen molecule. In order such interaction could occur, the plant and bacteria of a certain genotype should meet, i.e., a bacteria carrying the avirulence gene (avr) interacts with a plant, which has the corresponding R-gene. Such process is called an incompatible combination and leads to quick progressing of events, or to hypersensitive response. Receptors activate the passes of signal transduction and launch several protective systems. In the place of pathogen penetration, the strong oxidants are being synthesized such as H_2O_2, O_2 -, OH. Then the oxidative burst is being developed, followed by the death of infected cells according to the mechanism similar to apoptosis known in vertebrates. Hypersensitive response is being developed in the place of pathogen penetration into the neighboring cells. Rapidly developing local process produces signal molecules spreading along vascular system of a plant.

The scheme of molecular events occurring during pathogenesis of plants. SA - salicylic acid; JA - jasmonic acid; H_2O_2 - reactive oxygen species; PR - pathogenesis-related genes; SAR - systemic acquired resistance; ISR - induced systemic resistance; PCD - programmed cell death; PTGC - post-transcriptional gene silencing. By thin arrows are given direct involvement of a substance of gene product in a reaction. Empty arrows denote the events that need additional elements for signal transduction.

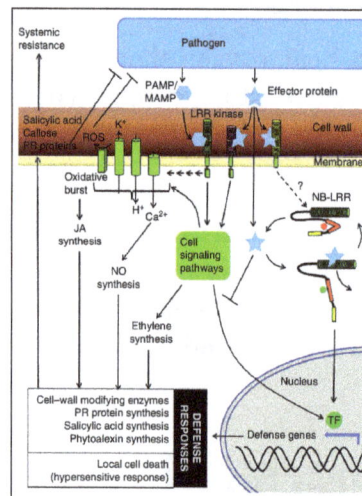

In dependence of local events, the set of signal molecules is being organized, which in turn, forms this or that generalized response. In the whole plant organism, the pathogenesis-related genes (PR-genes) are activated, the cell walls strengthen, and the plant accumulates some amount of protective substances, which are more effective in the struggle with this definite pathogenic form. In the plant cells, the salicylic acid (SA) is produced in considerable amounts and causes activation of SA-induced genes. The integrity of these events is named as systemic acquired resistance (SAR). The other systemic, or referring to the whole organism, response to pathogen infection is an induced systemic resistance (ISR). Its differs from SAR by activation of some differing set of pathogenesis genes and by other ways of signal transduction (without participance of salicylic acid).

In addition to immunity specific to pathogen, the general protective mechanisms are launched and, the plant becomes more resistant to other diseases after meeting the aviral pathogen.

The plant is not always supplied with receptors to the proteins of "attacking" bacteria or fungi. In this situation, the pathogen is called viral for given plant genotype; and the pair plant-pathogen is compatible. In this case, pathogen molecules are non-specific elicitors, which are non-specific substances causing pathogenesis. The ways of obtaining the signal from non-specific elicitors are still unknown. Various external stimuli (wound, non-specific elicitors) activate protein kinases and genes of signal molecules biosynthesis. In the course of signal transduction, the synthesis of JA, NO, H_2O_2, SA, and ethylene is produced. The processes are being activated that are well-known in animals: proteinkinase cascade, polyubiquitine-dependent protein degradation, etc. The signal transduction paths frequently intercross. For example, the gene of one of the key enzymes, PAL (phenylalanine ammonia-lyase), is activated not only in the course of hypersensitive response, but also in response to various external stimuli. PAL takes part in the synthesis of SA, phytoalexins, and lignin monomers.

The latter, both in hypersensitive response and non-specific pathogenesis induction, activate pathogenesis-related (PR) genes. To these genes are referring those encoding enzymes of protective substances biosynthesis, chitinases acting in degradation of fungi cell wall, enzymes of lignin biosynthesis, which is a component of plant cell wall, etc.

The positive and negative feedbacks stabilize the parameters of gene networks. For example, WRKY, transcription factor proceeding in PR-genes activation is regulated according to the principle of positive feedback, because it has the binding site in its own promoter. Along with H_2O_2 and salicylic acid biosynthesis, the events are developing, which lead to the death of infected cells. The cell dies together with the source of signaling molecules. Hence, the gene network terminates its functioning by the mechanism of the negative feedback.

The disease cycle of Xf, the bacterial pathogen that causes PD, involves intriguing interactions with plant and insect hosts. The bacterium persists and multiplies in both types of hosts. In plants, colonization is limited to the xylem. Several species of xylem-feeding insects, predominantly leaf-hoppers but also spittlebugs, can transmit Xf while feeding on host plants. Juvenile insects can transmit the pathogen until they molt and adult insects transmit Xf throughout their lifespans. After entering the plant, the pathogen multiplies, forming microcolonies at the inoculation site. The bacteria then efficiently and systemically colonize the plant by moving within and between xylem vessels. Movement of the pathogen between vessels is correlated with the expression of pathogen genes encoding degradative enzymes that are predicted to facilitate movement by degrading the pit membranes between vessels. Although the pathogen spreads through the plant and is detected in low numbers in many vessels, symptoms do not appear unless vessels contain high populations of bacteria. Xf is not observed outside of the plant's vascular system. Recent genome sequence data and comparative analyses with sequences of other bacterial pathogens show that Xf colonizes insect and plant hosts and induces disease in the plant host using genes that are expressed from a relatively small (2.5 Mb) genome.

The complete genome sequence of Xf (including sequences for two strains and draft sequences for another two) and the microarrays based on the sequence are facilitating the identification of genes that could lead to effective management strategies. In fact, many genes have been implicated or eliminated from consideration based on comparison of the Xf genome sequences with other pathogenic bacteria. For example, genes encoding the type III secretion system, which are common to and essential for virulence in many mammalian and plant pathogenic bacteria, are not found in Xf.

Several genes similar to those encoding putative virulence factors were identified in Xf. The presence of sequences related to virulence genes found in other organisms can inform the process of creating hypotheses; it does not prove that the genes are involved in pathogenicity. To determine gene function, in this case in the induction of disease, requires systematic analysis of mutagenesis and complementation, gene expression, physiologic and biochemical activity, and pathogen–host plant interactions. Several hypotheses that describe putative roles of particular genes in disease have been built from comparisons of genome sequences. Targeting genes for functional analysis will allow critical questions related to disease to be addressed.

Cell Attachment

Is attachment to cells in the insect vector or host plant critical for transmission and induction of disease? Genomic analyses indicate that Xf strains have genes related to those that encode hemagglutinins, adhesions, sticking pili, and fimbriae that could mediate different attachment strategies in the insect gut and plant xylem vessels. Attachment of bacteria to host tissues is important for colonization and pathogenicity for several pathogenic bacteria. Xf exhibits polar attachment to the insect cells and, although not polar, attachment to plant cells. Two genes, fimA and fimF, that encode major fimbrial proteins related to the Type I fimbriae of E. coli have been found in Xf. Site-directed mutants of the Xf fimA and fimF genes were developed to study their requirement for attachment in the development of virulence and disease. When they were grown in culture, the mutants, produced in two Xf strains from grapevines, were deficient in the two major fimbrial proteins and they exhibited reduced fimbriae size and number, cell aggregation, and cell size, compared with the parent strains. Both mutants remained pathogenic to grapevines, although their populations were slightly smaller than those of the wild-type strains. There is insufficient evidence to determine whether the mutations affect the insect transmission of Xf or whether they are important in attachment to the insect.

Targeted experiments guided by genome comparisons have led to the identification of genes that are essential to vector colonization. Xf contains a homologue of the Xanthomonas campestris pv. campestris rpfF gene, which is required for synthesis of a diffusible signaling factor (DSF) that regulates virulence. Mutational analysis of the Xf rpfF gene revealed that the gene also is required for production of DSF. Although the Xf rpfF mutant was not transmissible by insects, the mutants that were mechanically inoculated to plants were more virulent than were the wild-type Xf. Colonization of the insect foregut by the mutants was severely impaired, and it was associated with an inability of the bacterium to form biofilms in the insect. However, the mutant formed biofilms in the plant xylem. Why the rpfF mutants are more virulent to grapevine is not understood.

Motility and Disease

Is motility of Xf critical for disease? It is not clear whether Xf is motile within xylem and, if so, what significance motility would have for pathogenicity. The Xf genome does not contain sequences related to flagellar genes, and there is no evidence for flagellar motility. However, the genome sequence does contain sequences related to fimbrial and pili genes; there are at least three Type 4 fimbriae gene clusters in Xf–CVC and Xf–PD. Some of the genes associated with Type 4 pili, such as fimT-, pilZ-, and pilA-like genes, are present in more than one copy. Type 4 pili are involved in a sum of bacterial processes as adherence to surfaces, cell–cell interaction, twitching motility, and biofilm formation. Other Type 4 pili serve as receptors for bacteriophage and are required by some bacteria for natural transformation. Type 4 fimbriae also are involved in adhesion, surface translocation (twitching motility), and phase variation. It is not known whether products of those genes are involved in movement or adhesion of Xf within plant xylem vessels or the insect foregut.

Virulence Regulation

How are virulence genes regulated? The virulence genes of bacterial pathogens are frequently regulated by host signals, such as wound compounds or particular nutrient conditions, or through bacterial cell–cell signal molecules, such as those involved in quorum sensing. Xf contains genes that could be involved in quorum sensing and that could be important in sensing population size for activation of virulence genes. Gene array- and proteome-based studies demonstrate that activation of different gene networks is associated with different experimental growth conditions of Xf–CVC strain 9a5c.

Symptom Expression

What genes are involved in host symptom expression? Genome analyses have identified many genes that are similar to those that encode pathogenicity factors in other bacterial pathogens. Genes related to those for secreted toxins in other pathogenic microbes have been found in the Xf genome. The role of those genes in the scorch phenotype characteristic can now be explored. Other genes are similar to those that encode enzymes that produce xanthan gums, which could be involved in biofilm formation and vascular plugging, or that cause degradation of plant cell walls, allowing movement of bacteria between and within vessels. In addition to gum production and structural genes for fimbriae and pili, genome sequencing uncovered in Xf the presence of genes that encode large proteins similar to hemagglutinins (FhaB) from Bordetella pertussis and Neisseria meningiditis. FhaB protein is associated with Neisseria species that induce disease and in Bordetella, FhaB protein has been implicated in adhesion and tissue invasiveness. In another plant pathogen, Erwinia chrysanthemi, an FhaB homologue (HecA) contributes to the early stages of bacterial infection and epidermal cell death on the plant host. Like Xf strains, the plant pathogen Ralstonia solanacearum also carries multiple copies of FhaB homologues, however, none is as long as are those of Xf. The genes that encode those proteins in Xf could be as long as 10,000 bp, which is 10 times larger than the usual size of a bacterium gene. It is not known whether Xf FhaB homologue are responsible for disease symptoms.

The availability of the genome sequence for several Xf strains and for other bacterial pathogens and the development of techniques that permit mutation and complementation analyses of target genes in Xf, provide opportunities to systematically assess the roles of genes in pathogenicity. Studies are already identifying genes essential to pathogenicity that are potential targets for management strategies. To develop novel management strategies based on interference with bacterial pathogenicity, the committee makes the following recommendation.

Toward Host Plant Resistance

It is not well understood how the Xf–plant interaction results in disease symptoms, but in general, pathogens that target the xylem induce water stress in the host plant by increasing resistance to water flow. In fact several physiologic changes occur in grapevines infected with Xf that are also observed commonly in hosts infected with the pathogens that cause nonflaccid wilts, such as the verticillium wilts of hops, sunflower, and sycamore. In PD and nonflaccid wilts, the hosts develop marginal and interveinal chlorosis and necrosis with little or no wilting. The changes observed chlorosis, stomatal closure, membrane damage, lipid peroxidation, and increased superoxide anion accumulation are associated with plant senescence and resemble nondisease-induced water stress.

The xylem dysfunction caused by Xf has been attributed to the accumulation of bacterial polysaccharides or bacterial cell masses (colonies) that clog the elements or to host responses, such as the production of gels, gums, and tyloses is reduced water flow that is caused by clogged vessels; not increased cavitation and embolism of xylem elements. Reduced water flow also occurs in citrus plants affected with CVC. The blockage likely occurs as a result of the formation of large colonies of bacteria in the vessels, and it could be exacerbated by the formation of gels or gums. The plugging of vessels interferes with water flow through the leaf petioles to produce classic symptoms of water stress. The reduction of water flow by Xf inhibits plants' overnight recovery from daily transpirational water losses. Failure of the leaves to rehydrate makes the leaves more susceptible to damage from photoinhibition and high leaf temperatures. Prolonged water stress in diseased leaves, even if mild, can induce leaf senescence. Eventually the leaf dies and abscises. Symptom onset is accelerated in Xf-infected plants exposed to drought, because the stresses are additive.

Plant defense responses to wilt pathogens are common to defenses observed for other types of diseases. Those induced responses involve biosynthesis of structural barriers (including cell wall, callose, and lignin biosynthesis or modification), production of soluble inhibitory intermediates or compounds (such as active oxygen species or phytoalexins), and induced expression of genes for enzymes that degrade or otherwise affect the pathogen. Of importance to vascular diseases, those cellular precursors or products are transported into vessels, where they can alter the pathogen's environment. In response to many fungal vascular pathogens, callose materials are synthesized and secreted through the plasmalemma for final deposition onto the pits of the paravascular parenchyma cells at sites of attempted penetration and then onto the entire walls of infected vessels. Lignin precursors then polymerize onto the callose, producing a highly resistant barrier to degradation and penetration by pathogens. Inside the vessel lumen, other putative defense responses include the accumulation of phytoalexins; increased concentrations of phenolic compounds; and accumulation of enzymes, such as peroxidases, polyphenoloxidases, chitinases, and polygalacturonases. Collectively, those barriers and compounds can prevent or impede movement or growth across the vessel end walls, perforation plates, pit pairs between adjacent vessels, or the pits between vessels and xylem parenchyma. The barriers also can reduce pathogen-induced movement of nutrients from the xylem parenchyma cells into the vessel.

The formation of gums and tyloses in response to diseases caused by Xf was first reported by Esau, who observed that the formation of gums was among the first visible changes attributable to disease development. Excessive quantities of tyloses were found in the wood of vines with PD. Esau also noted the absence of a cork cambium and excessive accumulation of nonfunctional phloem in green patches on diseased canes. In contrast, normal cork cambium on healthy canes produced brown bark. Excess accumulation of gums and tyloses also was observed in Xf-infected peach with phony disease. However, only excessive gum accumulation occurred in Xf-infected alfalfa exhibiting symptoms of dwarf disease.

Differences between types of grapevines are apparent. Mollenhauer and Hopkins reported that the frequency of gums and tyloses is 7 or 8 times greater in tolerant muscadine (Vitis rotundifolia Mich) and wild grapevines than in those in susceptible bunch grapevines. They also observed gums and tyloses encapsulating the pathogen and speculated that the gums and tyloses imparted tolerance to PD in some grapevines. P.Y. Huang and colleagues observed the more extensive encapsulation of Xf cells in muscadine vines than in more susceptible bunch grapes and concluded from histochemical tests that the encapsulating materials were predominantly pectic substances of plant origin.Changes in the topography of the Xf cell wall from rippled to smooth were more prevalent in muscadine grapevines.

Although defense mechanisms are commonly observed in wilt diseases and have been observed in Xf–plant interactions, how they are activated, which processes are most effective in inhibiting pathogen spread, and what aspects can be manipulated to enhance resistance are currently unknown. The plant's responses are complex, and dissecting the relevance of each component in the process would be tedious and not necessarily revealing. Total genome expression profiling and other high-throughput technologies that allow analysis of complex interactions are being used to explain how those processes interact in disease and resistance for model plants whose genomes have been sequenced, such as Arabidopsis and rice. Because the genome sequence for grapevine is not available, application of information from model systems will allow development of hypotheses to explain disease and resistance in the Xf–grapevine interactions. However, the extent to which those model systems will apply is not clear.

A genome approach to analysis of grapevine is in its infancy, partially because the genome is so large (483 Mb over 38–40 chromosomes) and complexity (only 4% of the genome is transcribed). A multinational consortium recently announced an effort to sequence the complete Vitis genome. Efforts have been in place in several countries to develop molecular markers for grapevine and to construct and sequence cDNA libraries from various grapevine tissues in different stages of development or subjected to different stresses. There also is an effort to sequence cDNA libraries constructed from tissues of X –infected and uninfected grape genotypes that represent the susceptible V. vinifera (L.) and resistant and tolerant Vitis species (D. Cook, personal communication).

Host Plant Resistance

The use of resistant varieties is an environmentally acceptable and effective means of managing crop plant diseases if satisfactory, durable resistance can be incorporated into culturally desirable plants. Resistant varieties are particularly useful for protection against vascular pathogens that often cannot be adequately or reasonably controlled by other means. For example, the use of resistant varieties has stabilized rice production in areas where bacterial blight, caused by the vascular bacterial pathogen Xanthomonas oryzae pv. oryzae, is endemic.

Two kinds of resistance are recognized. Qualitative resistance, which confers specific resistance against some pathogen races, is easiest to incorporate into breeding programs because it is controlled by one or a few genes. However, qualitative resistance often is not durable because of changes in virulence within the pathogen population. Quantitative resistance, which is controlled by many genes, usually with small but additive effects, is considered more stable. However, quantitative traits for resistance have not been widely exploited for breeding programs because of the difficulty in accumulating multiple genes into plant varieties.

Quantitative and qualitative disease resistances have been described for grapevines. Qualitative resistance to Plasmopara viticola, the downy mildew pathogen, is controlled by a single dominant resistance gene that confers hypersensitivity at the point of infection. Multigenic quantitative resistance restricts the growth of mycelia beyond the site of penetration. Similarly, grapevines demonstrate both types of resistance to the powdery mildew pathogen (Uncinula necator) a qualitative type involving necrosis of the appressoria inside the epidermal cells and a quantitative type involving necrosis of host cells after the development of the fungal haustoria.

Genetic variation for resistance to Xf is available in several wild species of Vitis and in the genus Muscadina. Early studies identified three sources where inheritance of PD resistance was dominant (qualitative) and controlled by as many as three loci. The sources of resistance were three species of Vitis native to Florida; simpsoni, smalliana and shuttleworthi. Genetic sources of resistance vary greatly in the phenotypic expression of resistance, that is, in their ability to either resist or tolerate Xf. Tolerant vines grow well in the presence of the pest but permit pest build-up; resistant vines grow well in the presence of the pest but greatly limit pest populations. For example, some Vitis species, such as V. rotundifolia, V.shuttleworthi, and some selections of V. simpsonii Munson, restrict movement of Xf in the xylem; others, including V. aestivalis Michx and V. californica Benth., allow increased bacterial movement in the xylem.

Once a source of resistance is identified, it can be introgressed into agronomic varieties of a crop species through recurrent selection. However, the introduction of traits into grapevines, particularly wine grapes, is complicated by several factors. Excellent grape cultivars contain highly subtle combinations of many genes with small effects that can be disrupted by the sexual process. Good wine quality is traditionally associated with a few cultivars within V. vinifera; conservative consumers, producers, and sellers resist alteration of the genetic nature of those vines. Moreover, the wine business worldwide is built on the traditional wines of the classic European wine regions and the grape varieties historically associated with those regions. Wine is identified and marketed by those names; merlot, chardonnay, cabernet sauvignon, pinot grigio. Switching to a new variety with a new name, even if the variety does represent the solution to a production problem, is simply not a realistic answer for California wine, because the variety name is an integral part of the product identity. Clonal selection, which exploits genetic variation within traditional cultivars, has been used widely to improve grape cultivars. However, introduction of pest and disease resistance through intraspecific hybridization (crosses between genotypes of V. vinifera) has been used to improve table grapes and, in a few cases, wine grapes. Thus, development of hybrid wine grapes that are resistant to disease and pests is an important and feasible approach; however, hybridization with Vitis species other than V. vinifera is unacceptable in the global wine market because of the historical stigma attached to interspecific hybrid wine grapes and their subsequent prohibition in Europe.

Because the wine industry relies on a few select and very old cultivars for commercial production, classical breeding programs have not developed many new varieties that are commercially successful, and thus have not had a significant effect on grapevine improvement. However, those programs have significantly influenced the development of rootstock varieties that provide resistance to soil-borne pests and pathogens and to some abiotic problems. Rootstocks can contribute to PD management, and one advantage of their use for improving resistance is that a single new rootstock cultivar could be used with many wine grape varieties or clonal variants. Furthermore, breeding for traits in grape rootstocks, which are based on Vitis species other than V. vinifera, is much faster than is scion breeding because it can employ sources of resistance from those other species without requiring generations of recurrent selection to recover fruit quality. Those advantages have encouraged scientists to evaluate the effects of rootstock variety on PD symptoms and to consider the possibility that a rootstock could be developed that would produce mobile compounds that inhibit Xf or discourage GWSS feeding.

Novel Approaches to Controlling Xylella Fastidiosa

The hypersensitive response (HR) is a defense that can contribute to inhibition of infection by vascular bacterial pathogens. Typically, the response occurs in one or a few cells at the site of pathogen invasion and can restrict pathogen growth in plant tissues. The hypersensitive response is under genetic control and shares characteristics of programmed cell death (PCD) in animals. Although it has not been described to occur in grapevine resistance to Xf, it does occur in defense responses to vascular pathogens of other crop species. In those cases, it is the xylem parenchyma, the living cells adjacent to the xylem, that exhibit the response. Control of leaf cell death and defense are linked in some cases. For example, the Arabidopsis lsd1 mutant and other lesion-mimic mutants show spontaneous cell death and broad-spectrum resistance. Identification of the controls of PCD, therefore, could lead to novel strategies for disease control. Expression of the genes that inhibit PCD reveals that it is important for disease and for resistance, depending on the host–pathogen interaction. Expression of PCD inhibitors allows accessibility to a biotrophic fungal pathogen of barley and yet reduces disease symptoms caused by necrotrophic pathogens. Thus, altering PCD in plants is not a short-term strategy for disease control in any plant–pathogen system. Significant research is needed to identify plant pathways to PCD and to determine the repercussions of manipulating that essential cellular response.

Transgenic Approaches to Resistance

Transgenic technology for introducing traits for resistance to pathogens and pests shows promise because it could bypass the many complications and time constraints that attend the induction of resistance through breeding. If only one or a few genes are introduced, the use of transgenic technology could alleviate concerns about significant genetic changes in the grape variety but still allow for improvement of disease and pest resistance, productivity, and wine quality.

Efficient transformation systems that are applicable to a wide range of cultivars are key to the successful application of transgenic technologies. One major barrier has been the inability to regenerate plants from transformed tissues. With the use of embryonic cell lines as target tissues for transformation, many laboratories are now efficiently transforming and regenerating grapevine plants. The next limitation is that there are not many genes with known function that could be introduced to target desired effects. However, progress is being made. Of relevance to PD, antimicrobial peptides (such as lytic peptide, Shiva-I, defensins, and polygalacturonase-inhibiting proteins) are being introduced into the major wine grape cultivars of V. vinifera, and those transgenic lines should be tested for improved resistance to Xf.

Even with the tremendous advances that transgenic technologies have made in the potential for grapevine improvement, there are huge hurdles that are unrelated to the science of transgenics that must be overcome before technology can help solve the PD–GWSS problem. Vivier and Pretorius discuss several areas that remain beyond the science, including legal and regulatory issues, intellectual property and patenting, political and economic barriers, problems with marketing, traditional and cultural objections, and public perception. Public perception is particularly relevant to California's current sociological landscape, which is not receptive to the use of genetically modified plants. Although not insurmountable, the successful commercialization of grapevine varieties improved through transgenic technologies will depend on the resolution of those areas of difficulty.

The assumption of introducing useful genes and minimizing disruption of desirable complex trait combinations currently is reasonable but can only be predicted, not assured. Thousands of transgenic plants are likely to be discarded before one is developed that has the right combination of traits. Nevertheless, the information gained in the process of using powerful transformation tools can provide valuable insights to the basis of plant resistance.

Viral Diseases of Plants

Tobacco Mosaic

This plant disease is caused by Tobacco Mosaic Virus (TMV). It is known to occur in all tobacco growing countries of the world. It was first described in detail by Adolph Mayer in 1886. Tobacco Mosaic Virus specially infects tobacco and members of family Solanaceae. The virus may affect more than 150 genera of herbaceous dicotyledonous plants including many vegetables, flowers and weeds.

Symptoms

It produces mosaic like symptoms on plants. The symptoms on the healthy plant appear after ten days of infection. The first symptom is light green coloration between the veins of the young leaves. This is quickly followed by the development of mosaic or mottled pattern of light and dark green areas in the leaves. Leaves on infected plants are often small, curled and puckered. Symptoms on plants include chlorosis, curling, mottling, dwarfing, distortion and blistering of leaves. Mosaic does not result in plant death, but if infection occur early in the season, plants are stunted.

Etiology

Causal Organism

- Genus: Tobamovirus symptoms.
- Species: TMV.

- Species TMV named after Tobacco Mosaic Virus.

It is the first virus to be discovered by Iwanowski in 1892. It is a rod shaped virus. The virion measures 18 x 300 nanometers and weighs 39 million Daltons. The capsid is made from 2130 protein subunits (each subunit consists of 158 amino acids) and one molecule of genomic single stranded RNA. The ssRNA consists of 6,400 nucleotides.

Cycle

It is the most persistent plant virus. It has been known to survive 50 years in dried plant parts. The most common sources of virus inoculum for TMV are the debris of infected plants that remain in the soil and certain infected tobacco products which contaminate workers hand.

It is also transmitted mechanically by vegetative propagation of plants, grafting, seeds, pollens and being carried on the mouth parts of chewing insects. Once the virus enters the host, it begins to multiply by inducing cells to form new virus.

It takes over the metabolic cell processes resulting in abnormal cell function. Its RNA direct the synthesis of viral protein. After the RNA and the protein is produced, they undergo self assembly. After the assembly they are released outside the cell after its lysis.

Control Measures

There are no known chemical treatment used under field conditions that eliminate viral infection from plant tissues once it occurs.

However, some important control measures to check the infection are:

i.   Discarding infected plants.

ii. Growing virus free plants.

iii. Propagate plants via seeds rather than vegetatively.

iv. Crop rotation.

v.   Growing of resistant strains.

vi. Removing of all weeds during and after the growing season.

vii. Disinfecting tools by placing them in boiling water for 5 minutes and washing with a strong soap or detergent solution.

viii. Discouraging use of tobacco by workers.

ix. Encouraging the practice of washing hands by workers with soap and water before and after handling plants.

Cucumber Mosaic

This disease is known to occur world wide in both temperate and tropical regions. The virus was first found in cucurbits (cucumis sativus) showing mosaic symptoms and since then it is called as cucumber mosaic. The Cucumber Mosaic Virus (CMV) has a wide range of hosts. It attacks approximately 1200 species of 100 families and causes significant losses in many vegetable and horticultural crops.

Symptoms

Symptoms occur when the plants are about 6 weeks old and growing vigorously. Seedlings are seldom attacked. Four to five days after infection, the young leaves become mottled, distorted, wrinkled and then edges begin to curl downward forming rossette like clump near the ground (dwarfed appearance).

Older leaves develop chlorotic and then necrotic areas along the margins which spread laterally over the entire leaf. Fruits (cucumbers) develop pale green or white areas intermingled with green bumpy areas (often called white pickle). Infected fruits are oddly shaped and appear gray. They often have bitter taste and after picking become soft and spongy.

Etiology

Causal organism

Genus: Cucumovirus

Species: Cucumber Mosaic Virus

It is a linear positive sense single stranded RNA (ssRNA) virus. The CMV genome consists of three viral strands messenger-sense RNA molecules designated as RNA1, RNA2 and RNA3. Each RNA is enclosed with a protective protein coat with each being a distinct, single, spherical shaped particle.

Cycle

Perennial, biennial and annual weeds harbour CMV in roots, tubers and underground parts. These are important sources of virus transmission. The virus is also transmitted from infected plants to healthy plants by aphids (Aphis gossyii, Myzus persicae etc.), mechanical methods, sap, seeds, clothes and hands of the workers. It enters the host cell and replicates in the cytoplasm. The movement through cell to cell occurs by plasmodesmata or phloem.

Control Measures

1. Development of genetic resistance in cucumber species.

2. Eradicate of infected plants and perennial weeds.

3. Using trap crop method (growing resistant varieties around the perimeter of the field and to place the susceptible plants in the middle).

4. Use of insecticides to control aphid population.

5. Crop rotation.

Yellow Vein Mosaic of Bhindi

Yellow vein mosaic of bhindi (Abelomoschus esculents) or vein clearing of bhindi most devastating disease in all the bhindi growing regions of India. In case the plants get infected at early stages of development it causes 80% of crop loss.

Symptoms

The diseased plants can be recognised from a distance due to the yellowing of entire network of veins. The characteristic symptoms of the disease are the homologous network of yellow veins enclosing islands of green tissue within. In severe cases entire leaf become chlorotic.

Infected plants stunted and bear very deformed and small, yellow green fruits. Distortion of leaf stalks and stem occur at the advance stage of infection. The disease cause heavy loss in yield, if the plants get infected within 20 days after germination.

Etiology

Causal organism: Begomovirus and Bhindi yellow Vein Mosaic Virus.

Cycle

The disease is transmitted by white fly Bemisia tabaci. The population is high during hot summer months, the crop is seriously affected then. The virus also survives on various weeds growing along the roadside for e.g. Croton sparsifolia, Ageratum etc.

Control Measures

i. The vectors that are responsible for the spread of virus need to be controlled by spraying dimethoate 0.03 percent or monocrotophos 0.05 at 10 days intervals, (spraying must be done at late hours).

ii. Foliar spray of 5 to 10 ml neem oil in a litre of water at weekly intervals.

iii. Removing and destroying disease affected plants from crop fields to avoid secondary spread.

iv. To destroy the host weeds such as croton.

v. Crop rotation.

vi. Use seeds collected from disease free plants.

vii. Growing resistant varieties for e.g. Akra anamia, Akra Abhay. Punjab Padmini etc.

Little Leaf of Brinjal

This disease of brinjal was first reported in India in 1938. It was first considered a disease caused by virus. However, in 1969, it was attributed to a Mycoplasma-like organism. As far as known this disease occurs in India and Sri Lanka only. It is a serious disease of brinjal and can cause 100% yield loss in diseased plants.

Symptoms

The characteristic symptom of the disease is the smallness of the leaves. Petioles get shortened. Lamina of the leaves becomes soft and pale yellow. Internodes of the top branches are shortened resulting in a bushy appearance of the affected plant. Mostly there is no flowering but if flowers are formed they remain green. Fruiting is rare in affected plants. However, if any fruit is formed, it becomes hard, tough and fails to mature.

Etiology

Pathogen Phytoplasma

Mycoplasma like bodies (MLB) have been detected in phloem cells of roots of brinjal plant affected with little leaf disease. The size of MLBs varies from 230 nm to 770 nm. Each MLB contains ribosomes and nuclear material suspended by 16.5 nm wide tripple layered unit membrane.

Cycle

It is a sap transmissible disease. It occurs in nature of Datura fastuosa, Vinca rosea etc. The disease is transmitted by the insects (leaf hoppers), Cestius phycytis (Eutettix phycites) and Empoosco devostolls. The pathogen also survives on weed hosts.

Control Measures

1. Fort-nightly spray of the insecticides such as Ekalox or Folidol till the fruit set helps to check the spread of disease.

2. The insect vectors can also be controlled by spraying the crop by Dimethoate (Rogor-30 EC) or Oxydemiton methyl (Metasystox 25 EG) or Monocrotophos (monocil) @ 1 ml per litre of water.

3. Disease can be controlled when phorate @ 1.0 kg/ha is applied to the seed beds and seedlings are dipped in acqueous solution of 0.05% tetracyline along with 0.05% mono-crotophos.

4. Growing of disease resistant varieties such as Brinjal round, Black beauty, Pusa purple cluster etc.

5. Use of barriers of trap crops.

6. Sowing can be adjusted to avoid main flights of the insects.

7. Crop rotation.

8. Early removal of infected plants.

Barley Yellow Dwarf

The barley yellow dwarf virus infects several grains and staple crops, including wheat. Aphids primarily spread the virus. The virus causes discoloration of leaves and the tips of the plants, which reduces photosynthesis, stunts growth and decreases production of seed grains.

Wheat field

Bud Blight

Soybean field

The bud blight virus infects soybeans, a staple crop. It causes the stem to bend at the top and the buds to turn brown and drop off the plant. Nematodes spread this virus.

Sugarcane Mosaic Virus

Sugarcane field

The sugarcane mosaic virus discolors leaves of the sugarcane plant, restricting its ability to feed itself through photosynthesis and grow. It stunts the growth of young plants. Aphids and infected seed spread the virus.

Cauliflower Mosaic Virus

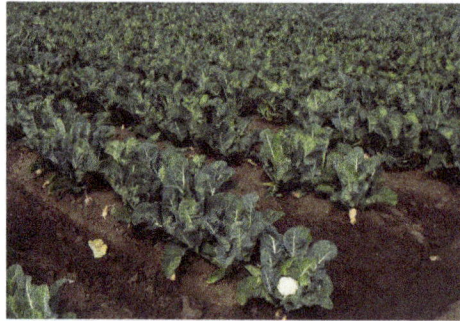

Cauliflower field

The cauliflower mosaic virus infects members of the brassica, or mustard, family, which includes cabbage, brussels sprouts, cauliflower, broccoli and rape seed. It causes a mosaic or mottle on the leaves, which stunts growth. Aphids and mechanical exposure spread the virus.

Lettuce Mosaic Virus

Lettuce crops

The lettuce mosaic virus mottles the leaves of almost all types of lettuce, stunting its growth and eliminating its market appeal. Aphids and infected seeds spread the virus.

Maise Mosaic Virus

Corn field

The maise mosaic virus causes yellow spots and stripes on the leaves of corn, stunting its growth. Leafhoppers spread the virus.

Peanut Stunt Virus

Peanuts

The peanut stunt virus causes discoloration and distortion of the leaves of peanuts and some other rhizomes, stunting their growth. Aphids and sap spread the virus.

Biological Control of Plant Pathogens

Plant diseases need to be controlled to maintain the quality and abundance of food, feed, and fiber produced by growers around the world. Different approaches may be used to prevent, mitigate or control plant diseases. Beyond good agronomic and horticultural practices, growers often rely heavily on chemical fertilizers and pesticides. Such inputs to agriculture have contributed significantly to the spectacular improvements in crop productivity and quality over the past 100 years. However, the environmental pollution caused by excessive use and misuse of agrochemicals, as well as fear-mongering by some opponents of pesticides, has led to considerable changes in people's attitudes towards the use of pesticides in agriculture. Today, there are strict regulations on chemical pesticide use, and there is political pressure to remove the most hazardous chemicals from the market. Additionally, the spread of plant diseases in natural ecosystems may preclude successful application of chemicals, because of the scale to which such applications might have to be applied. Consequently, some pest management researchers have focused their efforts on developing alternative inputs to synthetic chemicals for controlling pests and diseases. Among these alternatives are those referred to as biological controls.

A variety of biological controls are available for use, but further development and effective adoption will require a greater understanding of the complex interactions among plants, people, and the environment.

Types of Interactions Contributing to Biological Control

Throughout their lifecycle, plants and pathogens interact with a wide variety of organisms. These interactions can significantly affect plant health in various ways. In order to understand the mechanisms of biological control, it is helpful to appreciate the different ways that organisms interact. In order to interact, organisms must have some form of direct or indirect contact. Odum proposed that the interactions of two populations be defined by the outcomes for each. The types of interactions were referred to as mutualism, protocooperation, commensalism, neutralism, competition, amensalism, parasitism, and predation. While the terminology was developed for macroecology, examples of all of these types of interactions can be found in the natural world at both the macro-scopic and microscopic level. And, because the development of plant diseases involves both plants and microbes, the interactions that lead to biological control take place at multiple levels of scale.

From the plant's perspective, biological control can be considered a net positive result arising from a variety of specific and non-specific interactions. Using the spectrum of Odum's concepts, we can begin to classify and functionally delineate the diverse components of ecosystems that contribute to biocontrol. Mutualism is an association between two or more species where both species derive benefit. Sometimes, it is an obligatory lifelong interaction involving close physical and biochemical contact, such as those between plants and mycorrhizal fungi. However, they are generally facultative and opportunistic. For example, bacteria in the genus Rhizobium can reproduce either in the soil or, to a much greater degree, through their mutualistic association with legume plants. These types of mutualism can contribute to biological control, by fortifying the plant with improved nutrition and/ or by stimulating host defenses. Protocooperation is a form of mutualism, but the organisms involved do not depend exclusively on each other for survival. Many of the microbes isolated and classified as BCAs can be considered facultative mutualists involved in protocooperation, because survival rarely depends on any specific host and disease suppression will vary depending on the prevailing environ-mental conditions. Further down the spectrum, commensalism is a symbiotic interaction between two living organisms, where one organism benefits and the other is neither harmed nor benefited.

Most plant-associated microbes are assumed to be commensals with regards to the host plant, because their presence, individually or in total, rarely results in overtly positive or negative consequences to the plant. And, while their presence may present a variety of challenges to an infecting pathogen, an absence of measurable decrease in pathogen infection or disease severity is indicative of commensal in-teractions. Neutralism describes the biological interactions when the population density of one species has absolutely no effect whatsoever on the other. Related to biological control, an inability to associate the population dynamics of pathogen with that of another organism would indicate neutralism.

In contrast, antagonism between organisms results in a negative outcome for one or both. Com-petition within and between species results in decreased growth, activity and/or fecundity of the interacting organisms. Biocontrol can occur when non-pathogens compete with pathogens for nu-trients in and around the host plant. Direct interactions that benefit one population at the expense of another also affect our understanding of biological control. Parasitism is a symbiosis in which two phylogenetically unrelated organisms coexist over a prolonged period of time. In this type of association, one organism, usually the physically smaller of the two (called the parasite) benefits and the other (called the host) is harmed to some measurable extent. The activities of various hy-perparasites, i.e., those agents that parasitize plant pathogens, can result in biocontrol. And, inter-estingly, host infection and parasitism by relatively avirulent pathogens may lead to biocontrol of more virulent pathogens through the stimulation of host defense systems. Lastly, predation refers to the hunting and killing of one organism by another for consumption and sustenance. While the term predator typically refer to animals that feed at higher trophic levels in the macroscopic world, it has also been applied to the actions of microbes, e.g. protists, and mesofauna, e.g. fungal feed-ing nematodes and microarthropods, that consume pathogen biomass for sustenance. Biological control can result in varying degrees from all of these types of interactions, depending on the envi-ronmental context within which they occur. Significant biological control, as defined above, most generally arises from manipulating mutualisms between microbes and their plant hosts or from manipulating antagonisms between microbes and pathogens.

Mechanisms of Biological Control

Because biological control can result from many different types of interactions between organisms, researchers have focused on characterizing the mechanisms operating in different experimental situations. In all cases, pathogens are antagonized by the presence and activities of other organisms that they encounter. Here, we assert that the different mechanisms of antagonism occur across a spectrum of directionality related to the amount of interspecies contact and specificity of the interactions. Direct antagonism results from physical contact and/or a high-degree of selectivity for the pathogen by the mechanism(s) expressed by the BCA(s). In such a scheme, hyperparasitism by obligate parasites of a plant pathogen would be considered the most direct type of antagonism because the activities of no other organism would be required to exert a suppressive effect. In contrast, indirect antagonisms result from activities that do not involve sensing or targeting a pathogen by the BCA(s). Stimulation of plant host defense pathways by non-pathogenic BCAs is the most indirect form of antagonism. However, in the context of the natural environment, most described mechanisms of pathogen suppression will be modulated by the relative occurrence of other organisms in addition to the pathogen. While many investigations have attempted to establish the importance of specific mechanisms of biocontrol to particular pathosystems, all of the mechanisms described below are likely to be operating to some extent in all natural and managed ecosystems. And, the most effective BCAs studied to date appear to antagonize pathogens using multiple mechanisms. For instance, pseudomonads known to produce the antibiotic 2,4-diacetylphloroglucinol (DAPG) may also induce host defenses. Additionally, DAPG-producers can aggressively colonize roots, a trait that might further contribute to their ability to suppress pathogen activity in the rhizosphere of wheat through competition for organic nutrients.

Table Types of interspecies antagonisms leading to biological control of plant pathogens.

Type	Mechanism	Examples
Direct antagonism	Hyperparasitism/predation	Lytic/some nonlytic mycoviruses *Ampelomyces quisqualis Lysobacter enzymogenes Pasteuria penetrans Trichoderma virens*
Mixed-path antagonism	Antibiotics	2,4-diacetylphloroglucinol Phenazines Cyclic lipopeptides
	Lytic enzymes	Chitinases Glucanases Proteases
	Unregulated waste products	Ammonia Carbon dioxide Hydrogen cyanide
	Physical/chemical interference	Blockage of soil pores Germination signals consumption Molecular cross-talk confused
Indirect antagonism	Competition	Exudates/leachates consumption Siderophore scavenging Physical niche occupation
	Induction of host resistance	Contact with fungal cell walls Detection of pathogen-associated, molecular patterns Phytohormone-mediated induction

Hyperparasites and Predation

In hyperparasitism, the pathogen is directly attacked by a specific BCA that kills it or its propagules. In general, there are four major classes of hyperparasites: obligate bacterial pathogens, hypoviruses, facultative parasites, and predators. Pasteuria penetrans is an obligate bacterial pathogen of root-knot nematodes that has been used as a BCA. Hypoviruses are hyperparasites. A classical example is the virus that infects Cryphonectria parasitica, a fungus causing chestnut blight, which causes hypovirulence, a reduction in disease-producing capacity of the pathogen. The phenomenon has controlled the chestnut blight in many places. However, the interaction of virus, fungus, tree, and environment determines the success or failure of hypovirulence. There are several fungal parasites of plant pathogens, including those that attack sclerotia (e.g. Coniothyrium minitans) while others attack living hyphae (e.g. Pythium oligandrum). And, a single fungal pathogen can be attacked by multiple hyperparasites. For example, Acremonium alternatum, Acrodontium crateriforme, Ampelomyces quisqualis, Cladosporium oxysporum, and Gliocladium virens are just a few of the fungi that have the capacity to parasitize powdery mildew pathogens. Other hyperparasites attack plant-pathogenic nematodes during different stages of their life cycles (e.g. Paecilomyces lilacinus and Dactylella oviparasitica). In contrast to hyperparasitism, microbial predation is more general and pathogen non-specific and generally provides less predictable levels of disease control. Some BCAs exhibit predatory behavior under nutrient-limited conditions. However,Trichoderma produce a range of enzymes that are directed against cell walls of fungi. However, when fresh bark is used in composts, Trichoderma spp. do not directly attack the plant pathogen, Rhizoctonia solani. But in decomposing bark, the concentration of readily available cellulose decreases and this activates the chitinase genes of Trichoderma spp., which in turn produce chitinase to parasitize R. solani.

Antibiotic-mediated Suppression

Antibiotics are microbial toxins that can, at low concentrations, poison or kill other microorganisms. Most microbes produce and secrete one or more compounds with antibiotic activity. In some instances, antibiotics produced by microorganisms have been shown to be particularly effective at suppressing plant pathogens and the diseases they cause. Some examples of antibiotics reported to be involved in plant pathogen suppression are listed in table below. In all cases, the antibiotics have been shown to be particularly effective at suppressing growth of the target pathogen in vitro and/or in situ. To be effective, antibiotics must be produced in sufficient quantities near the pathogen to result in a biocontrol effect. In situ production of antibiotics by several different biocontrol agents has been measured; however, the effective quantities are difficult to estimate because of the small quantities produced relative to the other, less toxic, organic compounds present in the phytosphere. And while methods have been developed to ascertain when and where biocontrol agents may produce antibiotics, detecting expression in the infection court is difficult because of the heterogenous distribution of plant-associated microbes and the potential sites of infection.

In a few cases, the relative importance of antibiotic production by biocontrol bacteria has been demonstrated, where one or more genes responsible for biosynthesis of the antibiotics have been manipulated. For example, mutant strains incapable of producing phenazines or phloroglucinols have been shown to be equally capable of colonizing the rhizosphere but much less capable of suppressing soilborne root diseases than the corresponding wild-type and complemented mutant strains. Several biocontrol strains are known to produce multiple antibiotics which can suppress one or more pathogens. For example, Bacillus cereus strain UW85 is known to produce both zwittermycin and kanosamine. The ability to produce multiple antibiotics probably helps to suppress diverse microbial competitors, some of which are likely to be plant pathogens. The ability to produce multiple classes of antibiotics, that differentially inhibit different pathogens, is likely to enhance biological control. More recently, Pseudomonas putida WCS358r strains genetically engineered to produce phenazine and DAPG displayed improved capacities to suppress plant diseases in field-grown wheat.

Table Some of antibiotics produced by BCAs

Antibiotic	Source	Target pathogen	Disease
2, 4-diace-tyl-phloroglu-cinol	*Pseudomonas fluorescens* F113	*Pythium* spp.	Damping off
Agrocin 84	*Agrobacterium radiobacter*	*Agrobacterium tumefaciens*	Crown gall
Bacillomycin D	*Bacillus subtilis* AU195	*Aspergillus flavus*	Aflatoxin con-tamination
Bacillomycin, fengycin	*Bacillus amyloliquefaciens-*FZB42	*Fusarium oxysporum*	Wilt
Xanthobaccin A	*Lysobacter* sp. strain SB-K88	*Aphanomyces cochlioides*	Damping off
Gliotoxin	*Trichoderma virens*	*Rhizoctonia solani*	Root rots
Herbicolin	*Pantoea agglomerans*C9-1	*Erwinia amylovora*	Fire blight
Iturin A	*B. subtilis* QST713	*Botrytis cinerea* and *R. solani*	Damping off
Mycosubtilin	*B. subtilis* BBG100	*Pythium aphanidermatum*	Damping off
Phenazines	*P. fluorescens* 2-79 and 30-84	*Gaeumannomyces gram-inis* var. *tritici*	Take-all
Pyoluteorin, pyrrolnitrin	*P. fluorescens* Pf-5	*Pythium ultimum* and *R. solani*	Damping off
Pyrrolnitrin, pseudane	*Burkholderia cepacia*	*R. solani* and *Pyricularia oryzae*	Damping off and rice blast
Zwittermicin A	*Bacillus cereus* UW85	*Phytophthora medicaginis* and *P. aphanidermatum*	Damping off

Lytic Enzymes and other by Products of Microbial Life

Diverse microorganisms secrete and excrete other metabolites that can interfere with pathogen growth and/or activities. Many microorganisms produce and release lytic enzymes that can hydro-lyze a wide variety of polymeric compounds, including chitin, proteins, cellulose, hemicellulose, and DNA. Expression and secretion of these enzymes by different microbes can sometimes result in the suppression of plant pathogen activities directly. For example, control of Sclerotium rolfsii by Serratia marcescens appeared to be mediated by chitinase expression. And, a b-1,3-glucanase contributes significantly to biocontrol activities of Lysobacter enzymogenes strain C3. While they may stress and/or lyse cell walls of living organisms, these enzymes generally act to decompose plant residues and nonliving organic matter. Currently, it is unclear how much of the lytic en-zyme activity that can be detected in the natural environment represents specific responses to microbe-microbe interactions. It seems more likely that such activities are largely indicative of the need to degrade complex polymers in order to obtain carbon nutrition. Nonetheless, microbes that show a preference for colonizing and lysing plant pathogens might be classified as biocontrol agents. Lysobacter and Myxobacteria are known to produce copious amounts of lytic enzymes, and some isolates have been shown to be effective at suppressing fungal plant pathogens. So, the lines between competition, hyperparasitism, and antibiosis are generally blurred. Furthermore, some products of lytic enzyme activity may contribute to indirect disease suppression. For example, oligosaccharides derived from fungal cell walls are known to be potent inducers of plant host defenses. Interestingly, Lysobacter enzymogenes strain C3 has been shown to induce plant host resistance to disease, though the precise activities leading to this induction are not entirely clear. The quantitative contribution of any and all of the above compounds to disease suppression is likely to be dependent on the composition and carbon to nitrogen ratio of the soil organic matter that serves as a food source for microbial populations in the soil and rhizosphere. However, such activities can be manipulated so as to result in greater disease suppression. For example, in post-harvest disease control, addition of chitosan can stimulate microbial degradation of pathogens similar to that of an applied hyperparasite. Chitosan is a non-toxic and biodegradable polymer of beta-1,4-glucosamine produced from chitin by alkaline deacylation. Amendment of plant growth substratum with chitosan suppressed the root rot caused by Fusarium oxysporum f. sp. radicis-lycopersici in tomato. Although the exact mechanism of action of chitosan is not fully understood, it has been observed that treatment with chitosan increased resistance to pathogens.

Other microbial byproducts also may contribute to pathogen suppression. Hydrogen cyanide (HCN) effectively blocks the cytochrome oxidase pathway and is highly toxic to all aerobic microorganisms at picomolar concentrations. The production of HCN by certain fluorescent pseudomonads is believed to be involved in the suppression of root pathogens. P. fluorescens CHAo produces antibiotics, siderophores and HCN, but suppression of black rot of tobacco caused by Thielaviopsis basicola appeared to be due primarily to HCN production. Howell et al. reported that volatile compounds such as ammonia produced by Enterobacter cloacac were involved in the suppression of Pythium ultimum-induced damping-off of cotton. While it is clear that biocontrol microbes can release many different compounds into their surrounding environment, the types and amounts produced in natural systems in the presence and absence of plant disease have not been well documented and this remains a frontier for discovery.

Competition

From a microbial perspective, soils and living plant surfaces are frequently nutrient limited environments. To successfully colonize the phytosphere, a microbe must effectively compete for the available nutrients. On plant surfaces, host-supplied nutrients include exudates, leachates, or senesced tissue. Additionally, nutrients can be obtained from waste products of other organisms such as insects (e.g. aphid honeydew on leaf surface) and the soil. While difficult to prove directly, much indirect evidence suggests that competition between pathogens and non-pathogens for nutrient resources is important for limiting disease incidence and severity. In general, soilborne pathogens, such as species of Fusarium and Pythium, that infect through mycelial contact are more susceptible to competition from other soil- and plant-associated microbes than those pathogens that germinate directly on plant surfaces and infect through appressoria and infection pegs. Genetic work of Anderson et al. revealed that production of a particular plant glycoprotein called agglutinin was correlated with potential of P. putida to colonize the root system. P. putida mutants deficient in this ability exhibited reduced capacity to colonize the rhizosphere and a corresponding reduction in Fusarium wilt suppression in cucumber. The most abundant nonpathogenic plant-associated microbes are generally thought to protect the plant by rapid colonization and thereby exhausting the limited available substrates so that none are available for pathogens to grow. For example, effective catabolism of nutrients in the spermosphere has been identified as a mechanism contributing to the suppression of Pythium ultimum by Enterobacter cloacae. At the same time, these microbes produce metabolites that suppress pathogens. These microbes colonize the sites where water and carbon-containing nutrients are most readily available, such as exit points of secondary roots, damaged epidermal cells, and nectaries and utilize the root mucilage.

Biocontrol based on competition for rare but essential micronutrients, such as iron, has also been examined. Iron is extremely limited in the rhizosphere, depending on soil pH. In highly oxidized and aerated soil, iron is present in ferric form, which is insoluble in water (pH 7.4) and the concentration may be as low as 10-18 M. This concentration is too low to support the growth of microorganisms, which generally need concentrations approaching 10-6 M. To survive in such an environment, organisms were found to secrete iron-binding ligands called siderophores having high affinity to sequester iron from the micro-environment. Almost all microorganisms produce siderophores, of either the catechol type or hydroxamate type. Kloepper et al. were the first to demonstrate the importance of siderophore production as a mechanism of biological control of Erwinia carotovora by several plant-growth-promoting Pseudomonas fluorescens strains A1, BK1, TL3B1 and B10. And, a direct correlation was established in vitro between siderophore synthesis in fluorescent pseudomonads and their capacity to inhibit germination of chlamydospores of F. oxysporum. As with the antibiotics, mutants incapable of producing some siderophores, such as pyoverdine, were reduced in their capacity to suppress different plant pathogens. The increased efficiency in iron uptake of the commensal microorganisms is thought to be a contributing factor to their ability to aggressively colonize plant roots and an aid to the displacement of the deleterious organisms from potential sites of infection.

Induction of Host Resistance

Plants actively respond to a variety of environmental stimuli, including gravity, light, temperature, physical stress, water and nutrient availability. Plants also respond to a variety of chemical stimuli produced by soil- and plant-associated microbes. Such stimuli can either induce or condition plant host defenses through biochemical changes that enhance resistance against subsequent infection by a variety of pathogens. Induction of host defenses can be local and/or systemic in nature, depending on the type, source, and amount of stimuli. Recently, phytopathologists have begun to characterize the determinants and pathways of induced resistance stimulated by biological control agents and other non-pathogenic microbes. The first of these pathways, termed systemic acquired resistance (SAR), is mediated by salicylic acid (SA), a compound which is frequently produced following pathogen infection and typically leads to the expression of pathogenesis-related (PR) proteins. These PR proteins include a variety of enzymes some of which may act directly to lyse invading cells, reinforce cell wall boundaries to resist infections, or induce localized cell death. A second phenotype, first referred to as induced systemic resistance (ISR), is mediated by jasmonic acid (JA) and/or ethylene, which are produced following applications of some nonpathogenic rhizobacteria. Interestingly, the SA- and JA- dependent defense pathways can be mutually antagonistic, and some bacterial pathogens take advantage of this to overcome the SAR. For example, pathogenic strains of Pseudomonassyringae produce coronatine, which is similar to JA, to overcome the SA-mediated pathway. Because the various host-resistance pathways can be activated to varying degrees by different microbes and insect feeding, it is plausible that multiple stimuli are constantly being received and processed by the plant. Thus, the magnitude and duration of host defense induction will likely vary over time. Only if induction can be controlled, i.e. by overwhelming or synergistically interacting with endogenous signals, will host resistance be increased.

Table Bacterial determinants and types of host resistance induced by biocontrol agents

Bacterial strain	Plant species	Bacterial determinant	Type	Reference
Bacillus mycoides strain Bac J	Sugar beet	Peroxidase, chitinase and β-1,3-glucanase	ISR	Bargabus et al. (2002)
Bacillus subtilis GB03 and IN937a	*Arabidopsis*	2,3-butanediol	ISR	Ryu et al. (2004)
Pseudomonas fluorescens strains				
CHA0	Tobacco	Siderophore	SAR	Maurhofer et al. (1994)
	Arabidopsis	Antibiotics (DAPG)	ISR	Iavicoli et al. (2003)
WCS374	Radish	Lipopolysaccharide	ISR	Leeman et al. (1995)
		Siderophore		Leeman et al. (1995)
		Iron regulated factor		Leeman et al. (1995)
WCS417	Carnation	Lipopolysaccharide	ISR	Van Peer and Schipper (1992)
	Radish	Lipopolysaccharide	ISR	Leeman et al. (1995)
		Iron regulated factor		Leeman et al. (1995)
	Arabidopsis	Lipopolysaccharide	ISR	Van Wees et al. (1997)
	Tomato	Lipopolysaccharide	ISR	Duijff et al. (1997)
Pseudomonas putida strains	*Arabidopsis*	Lipopolysaccharide	ISR	Meziane et al. (2005)
WCS 358	*Arabidopsis*	Lipopolysaccharide	ISR	Meziane et al. (2005)
		Siderophore	ISR	Meziane et al. (2005)
BTP1	Bean	Z,3-hexenal	ISR	Ongena et al. (2004)
Serratia marcescens 90-166	Cucumber	Siderophore	ISR	Press et al. (2001)

A number of strains of root-colonizing microbes have been identified as potential elicitors of plant host defenses. Some biocontrol strains of Pseudomonas sp. and Trichoderma sp. are known to strongly induce plant host defenses. In several instances, inoculations with plant-growth-promoting rhizobacteria (PGPR) were effective in controlling multiple diseases caused by different pathogens, including anthracnose (Colletotrichum lagenarium), angular leaf spot (Pseudomonas syringae pv. lachrymans and bacterial wilt (Erwinia tracheiphila). A number of chemical elicitors of SAR and ISR may be produced by the PGPR strains upon inoculation, including salicylic acid, siderophore, lipopolysaccharides, and 2,3-butanediol, and other volatile substances. Again, there may be multiple functions to such molecules blurring the lines between direct and indirect antagonisms. More generally, a substantial number of microbial products have been identified as elicitors of host defenses, indicating that host defenses are likely stimulated continually over the course of a plant's lifecycle. Excluding the components directly related to pathogenesis, these inducers include lipopolysaccharides and flagellin from Gram-negative bacteria; cold shock proteins of diverse bacteria; transglutaminase, elicitins, and β-glucans in Oomycetes; invertase in yeast; chitin and ergosterol in all fungi; and xylanase in Trichoderma. These data suggest that plants would detect the composition of their plant-associated microbial communities and respond to changes in the abundance, types, and localization of many different signals. The importance of such interactions is indicated by the fact that further induction of host resistance pathways, by chemical and microbiological inducers, is not always effective at improving plant health or productivity in the field.

Microbial Diversity and Disease Suppression

Plants are surrounded by diverse types of mesofauna and microbial organisms, some of which can contribute to biological control of plant diseases. Microbes that contribute most to disease control are most likely those that could be classified competitive saprophytes, facultative plant symbionts and facultative hyperparasites. These can generally survive on dead plant material, but they are able to colonize and express biocontrol activities while growing on plant tissues. A few, like avirulent Fusarium oxysporum and binucleate Rhizoctonia-like fungi, are phylogenetically very similar to plant pathogens but lack active virulence determinants for many of the plant hosts from which they can be recovered. Others, like Pythium oligandrum are currently classified as distinct species. However, most are phylogenetically distinct from pathogens and, most often, they are subspecies variants of the same microbial groups. Due to the ease with which they can be cultured, most biocontrol research has focused on a limited number of bacterial (Bacillus, Burkholderia, Lysobacter, Pantoea, Pseudomonas, and Streptomyces) and fungal (Ampelomyces, Coniothyrium, Dactylella, Gliocladium, Paecilomyces, and Trichoderma) genera. Still, other microbes that are more recalcitrant to in vitro culturing have been intensively studied. These include mycorrhizal fungi, e.g. Pisolithus and Glomus spp. that can limit subsequent infections, and some hyperparasites of plant pathogens, e.g. Pasteuria penetrans which attack root-knot nematodes. Because multiple infections can and do take place in field-grown plants, weakly virulent pathogens can contribute to the suppression of more virulent pathogens, via the induction of host defenses. Lastly, there are the many general micro- and meso-fauna predators, such as protists, collembola, mites, nematodes, annelids, and insect larvae whose activities can reduce pathogen biomass, but may also facilitate infection and/or stimulate plant host defenses by virtue of their own herbivorous activities.

While various epiphytes and endophytes may contribute to biological control, the ubiquity of mycorrhizae deserves special consideration. Mycorrhizae are formed as the result of mutualist symbioses between fungi and plants and occur on most plant species. Because they are formed early in the development of the plants, they represent nearly ubiquitous root colonists that assist plants with the uptake of nutrients (especially phosphorus and micronutrients). The vesicular arbuscular mycorrhizal fungi (VAM, also known as arbuscular mycorrhizal or endomycorrhizal fungi) are all members of the zygomycota and the current classification contains one order, the Glomales, encompassing six genera into which 149 species have been classified. Arbuscular mycorrhizae involve aseptate fungi and are named for characteristic structures like arbuscles and vesicles found in the root cortex. Arbuscules start to form by repeated dichotomous branching of fungal hyphae approximately two days after root penetration inside the root cortical cell.

Arbuscules are believed to be the site of communication between the host and the fungus. Vesicles are basically hyphal swellings in the root cortex that contain lipids and cytoplasm and act as storage organ of VAM. These structures may present intra- and inter-cellular and can often develop thick walls in older roots. These thick walled structures may function as propagules. During colonization, VAM fungi can prevent root infections by reducing the access sites and stimulating host defense. VAM fungi have been found to reduce the incidence of root-knot nematode. Various mechanisms also allow VAM fungi to increase a plant's stress tolerance. This includes the intricate network of fungal hyphae around the roots which block pathogen infections. Inoculation of apple-tree seedlings with the VAM fungi Glomus fasciculatum and G. macrocarpum suppressed apple replant disease caused by phytotoxic myxomycetes. VAM fungi protect the host plant against root-infecting pathogenic bacteria. The damage due to Pseudomonas syringae on tomato may be significantly reduced when the plants are well colonized by mycorrhizae. The mechanisms involved in these interactions include physical protection, chemical interactions and indirect effects. The other mechanisms employed by VAM fungi to indirectly suppress plant pathogens include enhanced nutrition to plants; morphological changes in the root by increased lignification; changes in the chemical composition of the plant tissues like antifungal chitinase, isoflavonoids, etc.; alleviation of abiotic stress and changes in the microbial composition in the mycorrhizosphere. In contrast to VAM fungi, ectomycorrhizae proliferate outside the root surface and form a sheath around the root by the combination of mass of root and hyphae called a mantle. Disease protection by ectomycorrhizal fungi may involve multiple mechanisms including antibiosis, synthesis of fungistatic compounds by plant roots in response to mycorrhizal infection and a physical barrier of the fungal mantle around the plant root. Ectomycorrhizal fungi like Paxillus involutus effectively controlled root rot caused by Fusarium oxysporum and Fusarium moniliforme in red pine. Inoculation of sand pine with Pisolithus tinctorius, another ectomycorrhizal fungus, controlled disease caused by Phytophthora cinnamomi.

Because plant diseases may be suppressed by the activities of one or more plant-associated microbes, researchers have attempted to characterize the organisms involved in biological control. Historically, this has been done primarily through isolation, characterization, and application of individual organisms. By design, this approach focuses on specific forms of disease suppression. Specific suppression results from the activities of one or just a few microbial antagonists. This type of suppression is thought to be occurring when inoculation of a biocontrol agent results in substantial levels of disease suppressiveness. Its occurrence in natural systems may also occur from time to time. For example, the introduction of Pseudomonas fluorescens that produce the antibiotic 2,4-diacetylphloroglucinol can result in the suppression of various soilborne pathogens. However, specific agents must compete with other soil- and root-associated microbes to survive, propagate, and express their antagonistic potential during those times when the targeted pathogens pose an active threat to plant health. In contrast, general suppression is more frequently invoked to explain the reduced incidence or severity of plant diseases because the activities of multiple organisms can contribute to a reduction in disease pressure. High soil organic matter supports a large and diverse mass of microbes resulting in the availability of fewer ecological niches for which a pathogen competes. The extent of general suppression will vary substantially depending on the quantity and quality of organic matter present in a soil. Functional redundancy within different microbial communities allows for rapid depletion of the available soil nutrient pool under a large variety of conditions, before the pathogens can utilize them to proliferate and cause disease. For example, diverse seed-colonizing bacteria can consume nutrients that are released into the soil during germination thereby suppressing pathogen germination and growth. Manipulation of agricultural systems, through additions of composts, green manures and cover crops is aimed at improving endogenous levels of general suppression.

Biological Control of Fungal Plant Pathogens

Microorganisms naturally present in the plants ecosystem will help reduce disease potential or disease damage, but only if they are allowed to grow vigorously. They accomplish these tasks by competing with the pathogens for food sources, producing metabolites that inhibit the growth of the pathogens and physically eliminating the pathogens from the plant by occupying the space and sites first. Microorganisms not naturally present in plant environment can be introduced in an attempt to control diseases. This can be done by application of organic materials that contain natural microbial populations such as composts or natural microbial populations added to them including natural organic fertilizers with microbial supplements. In both cases, the products must be applied prior to disease development as they are preventive and not curative. Natural organic fertilizers should be used for their nutritional value (nitrogen and potassium) and not for any possible secondary effects.

Fungal plant pathogens are very diverse and cause diseases on different parts of plants such as root, stem, leaf, fruit, etc. In this topic, application of biological control strategies for controlling fungal diseases on different parts of plants will be discussed.

The majority of research on biocontrol of fungal diseases have focused on soil borne diseases rather than foliar or post harvest. According to the results of numerous research projects, several fungal and bacterial biocontrol agents have been used as seed and soil application to reduce the incidence of plant diseases caused by soil borne fungal pathogens. Since many plant pathogens can spread readily in the foliar parts, control of these diseases requires both suppression of initial plant infection and reduction of the infection rate. Granular applications of strain 1295-22 of Trihoderma harzianum has been shown to significantly inhibit disease severity of some plant diseases during the initial stage of disease development, most likely by reducing levels of the pathogen inoculum in the soil. It is apparent, therefore, that soil applications alone cannot effectively control the foliar phases of this disease.

Additives have been commonly used with fungicides to improve efficacy and they also may enhance the ability of biocontrol agents to reduce plant diseases. For example, it was reported that seed treatment using 10% Pelgel with solid matrix priming markedly enhanced the efficacy of Trichoderma strains to control Pythium sp. on various crops. Research has indicated that for control of multiple fungal plant diseases, greater control was obtained when Triton X-100 was included than when no additives, Pelgel, or Tween 20 were used. The use of specific surfactants with Trichoderma strains seems essential to obtain levels of control equivalent to those achieved with chemical fungicides. Detergents such as Triton X-100 may have several functions in biocontrol systems. They may slow the growth of pathogens more than that of the biocontrol agents or they may enhance wetting and adhesion of spores to infection courts. In preliminary experiments, both Tween 20 and Triton X-100 slowed the growth of both T. harzianum and the pathogens, but the ratio of the growth rates of T. harzianum and pathogens was greater with Triton X-100 than with Tween 20.

Living organisms, in addition to yielding a large quantity of biomass of the bioprotectant fungus, must perform effectively in each application. To examine this, different spore formulations of Trihoderma harzianum were compared in a study for controlling plant diseases. It was found that all formulations provided equivalent levels of control, indicating that the method of spore production may not be a key factor in the efficacy of this fungal biocontrol agent in controlling these diseases. To predictably and successfully use biological control agents for fungal disease control, it is critical that their biology and ecology be more completely understood. Therefore, effective antagonists must become established in plant ecosystems and remain active against target pathogens during periods favorable for plant infection.

Broadcast application of granules of Trichoderma to control plant diseases has resulted in establishment of stable and effective populations of plants in soils. Similarly, it was shown that the populations of T. harzianum in soils treated with spray applications were as high as those in soils treated with granular formulations. Population levels of strain 1295-22 in about 5 x 10 5 cfu g-1 of soil significantly reduced Pythium blight, root rot and brown patch diseases. However, spray applications, even though resulted in numerically similar levels of root colonization, did not provide the same benefit. This may reflect the differences in inoculum potential of granules versus spray applications. Granules are applied as a several-millimeter-diameter particle that is completely colonized by the fungus. Conidial inoculum, on the other hand, is much smaller and would therefore be expected to possess lower inoculum potential than the granular formulation.

Conversely, in greenhouse and field experiments, it was found that Trihoderma harzianum significantly reduced some foliar phases of plant diseases when spray applications of conidial suspensions containing Triton X-100 were used. Weekly spray applications were as effective as the standard (monthly) fungicide applications. These results indicate that the efficacy of T. harzianum against plant diseases, especially those involving secondary infections, is very strongly affected by the method of application

The ability to survive on the plant phylloplane is also a desirable trait for strains of fungal and bacterial antagonists used as biocontrol agents against foliar diseases. Spray applications of strain 1295-22 of T. harzianum has resulted in disease suppressive population levels on leaf. These populations were sufficient to suppress Pythium root rot, brown patch and dollar spot over the entire season. Thus, T. harzianum 1295-22 may possess a measure of phylloplane competence on the plants. The ideal biocontrol strategy attempts to introduce or promote the activity of biocontrol agents only when and where they are needed or are most effective and minimizes wasteful application of inoculum to non-target habitats. Thus, for effective delivery, it is necessary to consider plant–pathogen–antagonist interactions in terms of time and space.

Pythium, Rhizoctonia and Sclerotinia are important soil borne pathogens of many plant species and their survival structures in soil serve as primary inoculum. Consequently, suppression of the initial inoculum will be the first step in managing these pathogens. The granular application of biocontrol agents should be followed by monthly spray applications to suppress foliar phases of these diseases. Inhibition of the secondary infection and dissemination of these pathogens is also important for disease management. Monthly spray applications of T. harzianum could provide a second step in protection of plant foliage from attack by preventing these pathogens from initially infecting leaves and by reducing the spread of disease or other methods of inoculum dissemination. Finally, results of study have indicated that it will be necessary to apply weekly sprays for highly effective control of these pathogens under severe disease conditions.

In addition to Trihoderma and other fungal antagonists, several antagonistic bacterial species including Pseudomonas fluorescens, P. putida, P. aerofaciens, Burkholderia cepacia, Bacillus subtillis, B. Polymyxa and B. cerrues have also been used successfully in biological control of different soil borne fungal diseases. By application of these bacterial antagonists, various fungal pathogens including Rhizoctonia solani, Fusarium moxysporium, F. solani, Verticillium dahliae, Gaummannomuces graminis and soil borne diseases caused by them such as seed rot, damping-off, root rot, vascular wilt and take-all have been biologically controlled on major agricultural crops including cotton, sugar beet, wheat, rice and different vegetables.

Although the majority of biological control research have been concentrated on soil borne fungal diseases, a number of studies have focused on fungal pathogens causing diseases and disorders in above-ground parts of plants studied the possibility of biological control of fungal pathogens in the phylosphere and proposed that it may be possible to reduce the incidence and development of these diseases using fungal and bacterial antagonists.

In another study, biological control of powdery mildew disease on different crops using antagonistic fungi was investigated and it was found that biocontrol-active microorganisms can potentially be applied against this very important foliar diseases. Botrytis cinera which is the causal agent of gray mold on many plants was successfully controlled by the use of biocontrol-active microorganism on strawberry. In another study conducted by biological control of cotton leak of cucumber caused by a fungal foliar pathogen was studied. It was found that Bacillus cerrues, a bacterial antagonist was capable of reducing the incidence of the disease significantly.

Another example of using biocontrol-active microorganisms against foliar fungal pathogen is the study in which chestnut blight was successfully controlled by the virus that infects Cryphonectria parasitica, the fungal causal agent of the disease through the mechanism of hypovirulence, a reduction in pathogenicity of the pathogen. This phenomenon has resulted in control of the chestnut blight in many places. However, the interactions of virus, fungus, tree and environment play very important role in the success of disease control.

In addition to soil borne and foliar diseases some studies have also tested the efficacy of biocontrol-active microorganisms on post harvest fungal pathogens which cause losses to fruits and vegetables during post harvest and storage periods. Spray applications of fungal and bacterial antagonists have resulted in significant reduction in the infection caused by some fungal pathogens in the storage.

Biological Control of Nematodes

Nematodes are simple roundworms. Colorless, unsegmented, and lacking appendages, nematodes may be free-living, predaceous, or parasitic. Many of the parasitic species cause important diseases of plants, animals, and humans. Other species are beneficial in attacking insect pests, mostly sterilizing or otherwise debilitating their hosts. A very few cause insect death but these species tend to be difficult (e.g. tetradomatids) or expensive (e.g. mermithids) to mass produce, have narrow host specificity against pests of minor economic importance, possess modest virulence (e.g. sphaeruliids) or are otherwise poorly suited to exploit for pest control purposes. The only insect-parasitic nematodes possessing an optimal balance of biological control attributes are entomopathogenic or insecticidal nematodes in the genera Steinernema and Heterorhabditis. These multi-cellular metazoans occupy a biocontrol middle ground between microbial pathogens and predators/parasitoids, and are invariably lumped with pathogens, presumably because of their symbiotic relationship with bacteria.

Entomopathogenic nematodes are extraordinarily lethal to many important insect pests, yet are safe for plants and animals. This high degree of safety means that unlike chemicals, or even Bacillus thuringiensis, nematode applications do not require masks or other safety equipment; and re-entry time, residues, groundwater contamination, chemical trespass, and pollinators are not issues. Most biologicals require days or weeks to kill, yet nematodes, working with their symbiotic bacteria, can kill insects within 24-48 hours. Dozens of different insect pests are susceptible to infection, yet no adverse effects have been shown against beneficial insects or other nontargets in field studies. Nematodes are amenable to mass production and do not require specialized application equipment as they are compatible with standard agrochemical equipment, including various sprayers (e.g. backpack, pressurized, mist, electrostatic, fan, and aerial) and irrigation systems.

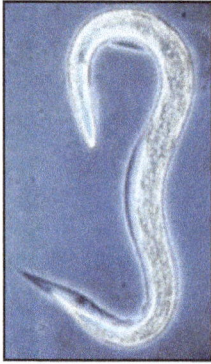

Entomopathogenic nematode infective juvenile

Infected caterpillar (wax moth larva) with nematodes emerging

Hundreds of researchers representing more than forty countries are working to develop nematodes as biological insecticides. Nematodes have been marketed on every continent except Antarctica for control of insect pests in high-value horticulture, agriculture, and home and garden niche markets.

Life Cycle

Steinernematids and heterorhabditids have similar life histories. The non-feeding, developmentally arrested infective juvenile seeks out insect hosts and initiates infections. When a host has been located, the nematodes penetrate into the insect body cavity, usually via natural body openings (mouth, anus, spiracles) or areas of thin cuticle. Once in the body cavity, a symbiotic bacterium (Xenorhabdus for steinernematids, Photorhabdus for heterorhabditids) is released from the nematode gut, which multiplies rapidly and causes rapid insect death. The nematodes feed upon the bacteria and liquefying host, and mature into adults. Steinernematid infective juveniles may become males or females, where as heterorhabditids develop into self-fertilizing hermaphrodites although subsequent generations within a host produce males and females as well.

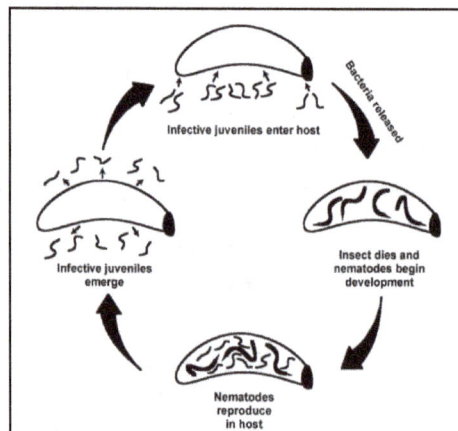

The life cycle is completed in a few days, and hundreds of thousands of new infective juveniles emerge in search of fresh hosts. Thus, entomopathogenic nematodes are a nematode-bacterium complex. The nematode may appear as little more than a biological syringe for its bacterial partner, yet the relationship between these organisms is one of classic mutualism. Nematode growth and reproduction depend upon conditions established in the host cadaver by the bacterium. The bacterium further contributes anti-immune proteins to assist the nematode in overcoming host defenses, and anti-microbials that suppress colonization of the cadaver by competing secondary invaders. Conversely, the bacterium lacks invasive powers and is dependent upon the nematode to locate and penetrate suitable hosts.

Production and Storage Technology

Entomopathogenic nematodes are mass produced for use as biopesticides using in vivo or in vitro methods. In vivo production (culture in live insect hosts) requires a low level of technology, has low startup costs, and resulting nematode quality is generally high, yet cost efficiency is low. The approach can be considered ideal for small markets. In vivo production may be improved through innovations in mechanization and streamlining. A novel alternative approach to in vivo methodology is production and application of nematodes in infected host cadavers; the cadavers (with nematodes developing inside) are distributed directly to the target site and pest suppression is subsequently achieved by the infective juveniles that emerge. In vitro solid culture, i.e., growing the nematodes on crumbled polyurethane foam, offers an intermediate level of technology and costs. In vitro liquid culture is the most cost-efficient production method but requires the largest startup capital. Liquid culture may be improved through progress in media development, nematode recovery, and bioreactor design. A variety of formulations have been developed to facilitate nematode storage and application including activated charcoal, alginate and polyacrylamide gels, baits, clay, paste, peat, polyurethane sponge, vermiculite, and water-dispersible granules. Depending on the formulation and nematode species, successful storage under refrigeration ranges from one to seven months. Optimum storage temperature for formulated nematodes varies according to species; generally, steinernematids tend to store best at 4-8 °C whereas heterorhabditids persist better at 10-15 °C.

Relative Effectiveness and Application Parameters

Growers will not adopt biological agents that do not provide efficacy comparable with standard chemical insecticides. Technological advances in nematode production, formulation, quality control, application timing and delivery, and particularly in selecting optimal target habitats and target pests, have narrowed the efficacy gap between chemical and nematode agents. Nematodes have consequently demonstrated efficacy in a number of agricultural and horticultural market segments.

Entomopathogenic nematodes are remarkably versatile in being useful against many soil and cryptic insect pests in diverse cropping systems, yet are clearly underutilized. Like other biological control agents, nematodes are constrained by being living organisms that require specific conditions to be effective. Thus, desiccation or ultraviolet light rapidly inactivates insecticidal nematodes; chemical insecticides are less constrained. Similarly, nematodes are effective within a narrower temperature range (generally between 20 °C and 30 °C) than chemicals, and are more impacted by suboptimal soil type, thatch depth, and irrigation frequency. Nematode-based insecticides may be inactivated if stored in hot vehicles, cannot be left in spray tanks for long periods, and are incompatible with several agricultural chemicals. Chemicals also have problems (e.g. mammalian toxicity, resistance, groundwater pollution, etc.) but a large knowledge base has been developed to support their use. Accelerated implementation of nematodes into IPM systems will require users to be more knowledgeable about how to use them effectively.

Therefore, based on the nematodes' biology, applications should be made in a manner that avoids direct sunlight, e.g. early morning or evening applications are often preferable. Soil in the treated area should be kept moist for at least two weeks after applications. Application to aboveground target areas is difficult due to the nematode's sensitivity to desiccation and UV radiation; however, some success against certain above-ground targets has been achieved and recently approaches have been enhanced by improved formulations. In all cases, the nematodes must be applied at a rate that is sufficient to kill the target pest; generally, 250,000 infective juveniles per m² of treated area is required (though in some cases an increased or slightly decreased rate may be suitable). Additionally, it is important to match the appropriate nematode species to the particular pest that is being targeted.

Appearance

Nematodes are formulated and applied as infective juveniles, the only free-living and therefore environmentally tolerant stage. Infective juveniles range from 0.4 to 1.5 mm in length and can be observed with a hand lens or microscope after separation from formulation materials. Disturbed nematodes move actively, however sedentary ambusher species (e.g. Steinernema carpocapsae, S. scapterisci) in water soon revert to a characteristic "J"-shaped resting position. Low temperature or oxygen levels will inhibit movement of even active cruiser species (e.g. S. glaseri, Heterorhabditis bacteriophora). In short, lack of movement is not always a sign of mortality; nematodes may have to be stimulated (e.g. probes, acetic acid, gentle heat) to move before assessing viability. Good quality nematodes tend to possess high lipid levels that provide a dense appearance, whereas nearly transparent nematodes are often active but possess low powers of infection.

Insects killed by most steinernematid nematodes become brown or tan, whereas insects killed by heterorhabditids become red and the tissues assume a gummy consistency. A dim luminescence given off by insects freshly killed by heterorhabditids is a foolproof diagnostic for this genus (the symbiotic bacteria provide the luminescence). Black cadavers with associated putrefaction indicate that the host was not killed by entomopathogenic species. Nematodes found within such cadavers tend to be free-living soil saprophages.

Habitat

Steinernematid and heterorhabditid nematodes are exclusively soil organisms. They are ubiquitous, having been isolated from every inhabited continent from a wide range of ecologically diverse soil habitats including cultivated fields, forests, grasslands, deserts, and even ocean beaches. When surveyed, entomopathogenic nematodes are recovered from 2% to 45% of sites sampled.

Pests Attacked

Because the symbiotic bacterium kills insects so quickly, there is no intimate host-parasite relationship as is characteristic for other insect-parasitic nematodes. Consequently, entomo-pathogenic nematodes are lethal to an extraordinarily broad range of insect pests in the laboratory. Field host range is considerably more restricted, with some species being quite narrow in host specificity. Nonetheless, when considered as a group of nearly 80 species, entomo-pathogenic nematodes are useful against a large number of insect pests. Additionally, entomo-pathogenic nematodes have been marketed for control of certain plant parasitic nematodes, though efficacy has been variable depending on species. A list of many of the insect pests that are commercially targeted with entomopathogenic nematodes is provided in the table below. As field research progresses and improved insect-nematode matches are made, this list is certain to expand.

Use of Nematodes as Biological Insecticides

Pest Common name	Pest Scientific name	Key Crop(s) targeted	Efficacious Nematodes *
Artichoke plume moth	*Platyptilia carduidactyla*	Artichoke	Sc
Armyworms	Lepidoptera: Noctuidae	Vegetables	Sc, Sf, Sr
Banana moth	*Opogona sachari*	Ornamentals	Hb, Sc
Banana root borer	*Cosmopolites sordidus*	Banana	Sc, Sf, Sg
Billbug	*Sphenophorus* spp. (Coleoptera: Curculionidae)	Turf	Hb,Sc
Black cutworm	*Agrotis ipsilon*	Turf, vegetables	Sc
Black vine weevil	*Otiorhynchus sulcatus*	Berries, ornamentals	Hb, Hd, Hm, Hmeg, Sc, Sg
Borers	*Synanthedon* spp. and other sesiids	Fruit trees & ornamentals	Hb, Sc, Sf
Cat flea	*Ctenocephalides felis*	Home yard, turf	Sc
Citrus root weevil	*Pachnaeus* spp. (Coleoptera: Curculionidae	Citrus, ornamentals	Sr, Hb
Codling moth	*Cydia pomonella*	Pome fruit	Sc, Sf
Corn earworm	*Helicoverpa zea*	Vegetables	Sc, Sf, Sr
Corn rootworm	*Diabrotica* spp.	Vegetables	Hb, Sc
Cranberry girdler	*Chrysoteuchia* topiaria	Cranberries	Sc
Crane fly	Diptera: Tipulidae	Turf	Sc
Diaprepes root weevil	*Diaprepes abbreviatus*	Citrus, ornamentals	Hb, Sr
Fungus gnats	Diptera: Sciaridae	Mushrooms, greenhouse	Sf, Hb
Grape root borer	*Vitacea polistiformis*	Grapes	Hz, Hb
Iris borer	*Macronoctua onusta*	Iris	Hb, Sc
Large pine weevil	*Hylobius albietis*	Forest plantings	Hd, Sc
Leafminers	*Liriomyza* spp. (Diptera: Agromyzidae)	Vegetables, ornamentals	Sc, Sf
Mole crickets	*Scapteriscus* spp.	Turf	Sc, Sr, Scap
Navel orangeworm	*Amyelois transitella*	Nut and fruit trees	Sc
Plum curculio	*Conotrachelus nenuphar*	Fruit trees	Sr

Scarab grubs**	Coleoptera: Scarabaeidae	Turf, ornamentals	Hb, Sc, Sg, Ss, Hz
Shore flies	*Scatella* spp.	Ornamentals	Sc, Sf
Strawberry root weevil	*Otiorhynchus ovatus*	Berries	Hm
Small hive beetle	*Aethina tumida*	Bee hives	Yes (Hi, Sr)
Sweetpotato weevil	*Cylas formicarius*	Sweet potato	Hb, Sc, Sf

Characteristics of Some Commercialized Species

1. Steinernema carpocapsae: This species is the most studied of all entomopathogenic nematodes. Important attributes include ease of mass production and ability to formulate in a partially desiccated state that provides several months of room-temperature shelf-life. S. carpocapsae is particularly effective against lepidopterous larvae, including various webworms, cutworms, armyworms, girdlers, some weevils, and wood-borers. This species is a classic sit-and-wait or "ambush" forager, standing on its tail in an upright position near the soil surface and attaching to passing hosts. Consequently, S. carpocapsae is especially effective when applied against highly mobile surface-adapted insects (though some below-ground insects are also controlled by this nematode). S. carpocapsae is also highly responsive to carbon dioxide once a host has been contacted, thus the spiracles are a key portal of host entry. It is most effective at temperatures ranging from 22 to 28 °C.

2. Steinernema feltiae: S. feltiae is especially effective against immature dipterous insects, including mushroom flies, fungus gnats, and tipulids as well some lepidopterous larvae. This nematode is unique in maintaining infectivity at soil temperatures as low as 10 °C. S. feltiae has an intermediate foraging strategy between the ambush and cruiser type.

3. Steinernema glaseri: One of the largest entomopathogenic nematode species at twice the length but eight times the volume of S. carpocapsae infective juveniles, S. glaseri is especially effective against coleopterous larvae, particularly scarabs. This species is a cruise forager, neither nictating nor attaching well to passing hosts, but highly mobile and responsive to long-range host volatiles. Thus, this nematode is best adapted to parasitize hosts possessing low mobility and residing within the soil profile. Field trials, particularly in Japan, have shown that S. glaseri can provide control of several scarab species. Large size, however, reduces yield, making this species significantly more expensive to produce than other species. A tendency to occasionally "lose" its bacterial symbiote is bothersome. Moreover, the highly active and robust infective juveniles are difficult to contain within formulations that rely on partial nematode dehydration. In short, additional technological advances are needed before this nematode is likely to see substantial use.

4. Steinernema kushidai: Only isolated so far from Japan and only known to parasitize scarab larvae, S. kushidai has been commercialized and marketed primarily in Asia.

5. Steinernema riobrave: This novel and highly pathogenic species was originally isolated from the Rio Grande Valley of Texas, but has since been also been isolated in other areas, e.g. in the southwestern USA. Its effective host range runs across multiple insect orders. This versatility is likely due in part to its ability to exploit aspects of both ambusher and cruiser means of finding hosts. Trials have demonstrated its effectiveness against corn earworm, mole crickets, and plum curculio. Steinernema riobrave has also been highly effective in suppressing citrus root weevils (e.g. Diaprepes abbreviates and Pachnaeus species). This nematode is active across a range of temperatures; it is effective at killing insects at soil temperatures above 35 °C, and can also infect at

15 °C. Persistence is excellent even under semi-arid conditions, a feature no doubt enhanced by the uniquely high lipid levels found in infective juveniles. Its small size provides high yields whether using in vivo (up to 375,000 infective juveniles per wax moth larvae) or in vitro methods.

6. Steinernema scapterisci: The only entomopathogenic nematode to be used in a classical biological control program, S. scapterisci was isolated from Uruguay and first released in Florida in 1985 to suppress an introduced pest, mole crickets. The nematode become established and presently contributes to control. Steinernema scapterisci is highly specific to mole crickets. Its ambusher approach to finding insects is ideally suited to the turfgrass tunneling habits of its host. Commercially available since 1993, this nematode is also sold as a biological insecticide, where its excellent ability to persist and provide long-term control contributes to overall efficacy.

7. Heterorhabditis bacteriophora: Among the most economically important entomopathogenic nematodes, H. bacteriophora possesses considerable versatility, attacking lepidopterous and coleopterous insect larvae, among other insects. This cruiser species appears quite useful against root weevils, particularly black vine weevil where it has provided consistently excellent results in containerized soil. A warm temperature nematode, H. bacteriophora shows reduced efficacy when soil drops below 20 °C.

8. Heterorhabditis indica: First discovered in India, this nematode is now known to be ubiquitous. Heterorhabditis indica is considered to be a heat tolerant nematode (infecting insects at 30 °C or higher). The nematode produces high yields in vivo and in vitro, but shelf life is generally shorter than most other nematode species.

9. Heterorhabditis megidis: First isolated in Ohio, this nematode is commercially available and marketed especially in western Europe for control of black vine weevil and various other soil insects. Heterorhabditis megidis is considered to be a cold tolerant nematode because it can effectively infect insects at temperatures below 15 °C.

Plant Growth-Promoting Bacteria for Biocontrol

Plant growth-promoting bacteria (PGPB) are associated with many, if not all, plant species and are commonly present in many environments. The most widely studied group of PGPB are plant growth-promoting rhizobacteria (PGPR) colonizing the root surfaces and the closely adhering soil interface, the rhizosphere. As reviewed by Kloepper et al. or, more recently, by Gray and Smith, some of these PGPR can also enter root interior and establish endophytic populations. Many of them are able to transcend the endodermis barrier, crossing from the root cortex to the vascular system, and subsequently thrive as endophytes in stem, leaves, tubers, and other organs. The extent of endophytic colonization of host plant organs and tissues reflects the ability of bacteria to selectively adapt to these specific ecological niches. Consequently, intimate associations between bacteria and host plants can be formed without harming the plant. Although, it is generally assumed that many bacterial endophyte communities are the product of a colonizing process initiated in the root zone, they may also originate from other source than the rhizosphere, such as the phyllosphere, the anthosphere, or the spermosphere.

Despite their different ecological niches, free-living rhizobacteria and endophytic bacteria use some of the same mechanisms to promote plant growth and control phytopathogens. The widely recognized mechanisms of biocontrol mediated by PGPB are competition for an ecological niche or a substrate, production of inhibitory allelochemicals, and induction of systemic resistance (ISR) in host plants to a broad spectrum of pathogens and/or abiotic stresses.

Competitive Root Colonization

Despite their potential as low-input practical agents of plant protection, application of PGPB has been hampered by inconsistent performance in field tests; this is usually attributed to their poor rhizosphere competence. Rhizosphere competence of biocontrol agents comprises effective root colonization combined with the ability to survive and proliferate along growing plant roots over a considerable time period, in the presence of the indigenous microflora. Given the importance of rhizosphere competence as a prerequisite of effective biological control, understanding root-microbe communication, as affected by genetic and environmental determinants in spatial and temporal contexts, will significantly contribute to improve the efficacy of these biocontrol agents.

Competition for root niches and bacterial determinants directly involves root colonization. The root surface and surrounding rhizosphere are significant carbon sinks. Photosynthate allocation to this zone can be as high as 40%. Thus, along root surfaces there are various suitable nutrient-rich niches attracting a great diversity of microorganisms, including phytopathogens. Competition for these nutrients and niches is a fundamental mechanism by which PGPB protect plants from phytopathogens. PGPB reach root surfaces by active motility facilitated by flagella and are guided by chemotactic responses. Known chemical attractants present in root exudates include organic acids, amino acids, and specific sugars. Some exudates can also be effective as antimicrobial agents and thus give ecological niche advantage to organisms that have adequate enzymatic machinery to detoxify them. The quantity and composition of chemoattractants and antimicrobials exuded by plant roots are under genetic and environmental control. This implies that PGPB competence highly depends either on their abilities to take advantage of a specific environment or on their abilities to adapt to changing conditions. As an example, Azospirillum chemotaxis is induced by sugars, amino acids, and organic acids, but the degree of chemotactic response to each of these compounds differs among strains. PGPB may be uniquely equipped to sense chemoattractants, e.g. rice exudates induce stronger chemotactic responses of endophytic bacteria than from non-PGPB present in the rice rhizosphere.

Bacterial lipopolysaccharides (LPS), in particular the O-antigen chain, can also contribute to root colonization. However, the importance of LPS in this colonization might be strain dependent since the O-antigenic side chain of Pseudomonas fluorescens WCS374 does not contribute to potato root adhesion, whereas the O-antigen chain of P. fluorescens PCL1205 is involved in tomato root colonization. Furthermore, the O-antigenic aspect of LPS does not contribute to rhizoplane colonization of tomato by the plant beneficial endophytic bacterium P. fluorescens WCS417r but, interestingly, this bacterial determinant was involved in endophytic colonization of roots.

It has also been recently demonstrated that the high bacterial growth rate and ability to synthesize vitamin B1 and exude NADH dehydrogenases contribute to plant colonization by PGPB. Another determinant of root colonization ability by bacteria is type IV pili, better known for its involvement in the adhesion of animal and human pathogenic bacteria to eukaryotic cells. The type IV pili also play a role in plant colonization by endophytic bacteria such as Azoarcus sp.

Root colonization and site-specific recombinase. Bacterial traits required for effective root colonization are subject to phase variation, a regulatory process for DNA rearrangements orchestrated by site-specific recombinase. In certain PGPB, efficient root colonization is linked to their ability to secrete a site-specific recombinase. Transfer of the site-specific recombinase gene from a rhizosphere-competent P. fluorescens into a rhizosphere-incompetent Pseudomonas strain enhanced its ability to colonize root tips.

Utilization of root exudates and root mucilage by PGPB.Since root exudates are the primary source of nutrients for rhizosphere microorganisms, rhizosphere competence implies that PGPB are well adapted to their utilization. Despite the fact that sugars have often been reported as the major carbon source in exudates, the ability to use specific sugars does not play a major role in tomato root colonization. Similarly, although amino acids are present in root exudates, the bioavailability of amino acids alone is considered insufficient to support root tip colonization by auxotrophic mutants of P. fluorescens WCS365. In contrast, Simons et al. reported that amino acid synthesis is required for root colonization by P. fluorescens WCS365, indicating that amino acid prototrophy is involved in rhizosphere competence. In addition, PGPB regulate the rate of uptake of polyamines such as putrescine, spermine, and spermidine, since their high titer could retard bacterial growth and reduce their ability to competitively colonize roots. Root mucilage also offers a utilizable carbon source for PGPB to use for the competitive colonization.

Biocontrol Activity Mediated by the Synthesis of Allelochemicals

Offensive PGPB colonization and defensive retention of rhizosphere niches are enabled by production of bacterial allelochemicals, including iron-chelating siderophores, antibiotics, biocidal volatiles, lytic enzymes, and detoxification enzymes.

Competition for iron and the role of siderophores.Iron is an essential growth element for all living organisms. The scarcity of bioavailable iron in soil habitats and on plant surfaces foments a furious competition. Under iron-limiting conditions PGPB produce low-molecular-weight compounds called siderophores to competitively acquire ferric ion. Although various bacterial siderophores differ in their abilities to sequester iron, in general, they deprive pathogenic fungi of this essential element since the fungal siderophores have lower affinity. Some PGPB strains go one step further and draw iron from heterologous siderophores produced by cohabiting microorganisms.

Siderophore biosynthesis is generally tightly regulated by iron-sensitive Fur proteins, the global regulators GacS and GacA, the sigma factors RpoS, PvdS, and FpvI, quorum-sensing autoinducers such as N-acyl homoserine lactone, and site-specific recombinases. However, some data demonstrate that none of these global regulators is involved in siderophore production. Neither GacS nor RpoS significantly affected the level of siderophores synthesized by Enterobacter cloacae CAL2 and UW4. RpoS is not involved in the regulation of siderophore production by Pseudomonas putida strain WCS358. In addition, GrrA/GrrS, but not GacS/GacA, are involved in siderophore synthesis regulation in Serratia plymuthica strain IC1270, suggesting that gene evolution occurred in the siderophore-producing bacteria. A myriad of environmental factors can also modulate siderophores synthesis, including pH, the level of iron and the form of iron ions, the presence of other trace elements, and an adequate supply of carbon, nitrogen, and phosphorus.

Antibiosis

The basis of antibiosis as a biocontrol mechanism of PGPB has become increasingly better understood over the past two decades. A variety of antibiotics have been identified, including compounds such as amphisin, 2,4-diacetylphloroglucinol (DAPG), hydrogen cyanide, oomycin A, phenazine, pyoluteorin, pyrrolnitrin, tensin, tropolone, and cyclic lipopeptides produced by pseudomonads and oligomycin A, kanosamine, zwittermicin A, and xanthobaccin produced by Bacillus, Streptomyces, and Stenotrophomonas spp. Interestingly, some antibiotics produced by PGPB are finding new uses as experimental pharmaceuticals, and this group of bacteria may offer an untapped resource for compounds to deal with the alarming ascent of multidrug-resistant human pathogenic bacteria.

Regulatory cascades of these antibiotics involve GacA/GacS or GrrA/GrrS, RpoD, and RpoS, N-acyl homoserine lactone derivatives and positive autoregulation. Antibiotic synthesis is tightly linked to the overall metabolic status of the cell, which in turn is dictated by nutrient availability and other environmental stimuli, such as major and minor minerals, type of carbon source and supply, pH, temperature, and other parameters. Trace elements, particularly zinc, and carbon source levels influence the genetic stability/instability of bacteria, affecting their ability to produce secondary metabolites. It is important to note that many strains produce pallet of secondary antimicrobial metabolites and that conditions favoring one compound may not favor another. Thus, the varied arsenal of biocontrol strains may enable antagonists to perform their ultimate objective of pathogen suppression under the widest range of environmental conditions. For example, in P. fluorescens CHA0 biosynthesis of DAPG is stimulated and pyoluteorin is repressed in the presence of glucose as a carbon source. As glucose is depleted, however, pyoluteorin becomes the more abundantly antimicrobial compound produced by this strain. This ensures a degree of flexibility for the antagonist when confronted with a different or a changeable environment. Biotic conditions can also influence antibiotic biosynthesis. For example bacterial metabolites salicylates and pyoluteorin can affect DAPG production by P. fluorescens CHA0. Furthermore, plant growth and development also influence antiobiotic production, since biological activity of DAPG producers is not induced by the exudates of young plant roots but is induced by the exudates of older plants, which results in selective pressure against other rhizosphere microorganisms. Plant host genotype also plays a significant role in the disease-suppressive interaction of plant with a microbial biocontrol agent, as demonstrated by Smith et al.

1. Lytic enzyme production: A variety of microorganisms also exhibit hyperparasitic activity, attacking pathogens by excreting cell wall hydrolases. Chitinase produced by S. plymuthica C48 inhibited spore germination and germ-tube elongation in Botrytis cinerea. The ability to produce extracellular chitinases is considered crucial for Serratia marcescens to act as antagonist against Sclerotium rolfsii, and for Paenibacillus sp. strain 300 and Streptomyces sp. strain 385 to suppress Fusarium oxysporum f. sp. cucumerinum. It has been also demonstrated that extracellular chitinase and laminarinase synthesized by Pseudomonas stutzeri digest and lyse mycelia of F. solani. Although, chitinolytic activity appears less essential for PGPB such as S. plymutica IC14 when used to suppress S. sclerotiorum and B. cinerea, synthesis of proteases and other biocontrol traits are involved. The β-1,3-glucanase synthesized by Paenibacillus sp. strain 300 and Streptomyces sp. strain 385 lyse

fungal cell walls of F. oxysporum f. sp. cucumerinum. B. cepacia synthesizes β-1,3-glucanase that destroys the integrity of R. solani, S. rolfsii, and Pythium ultimum cell walls. Similar to siderophores and antibiotics, regulation of lytic enzyme production (proteases and chitinases in particular) involves the GacA/GacS or GrrA/GrrS regulatory systems and colony phase variation.

2. Detoxification and degradation of virulence factors: Another mechanism of biological control is the detoxification of pathogen virulence factors. For example, certain biocontrol agents are able to detoxify albicidin toxin produced by Xanthomonas albilineans. The detoxification mechanisms include production of a protein that reversibly binds the toxin in both Klebsiella oxytoca and Alcaligenes denitrificans, as well as an irreversible detoxification of albicidin mediated by an esterase that occurs in Pantoea dispersa. Several different microorganisms, including strains of B. cepacia and Ralstonia solanacearum, can also hydrolyze fusaric acid, a phytotoxin produced by various Fusarium species. More often though, pathogen toxins display a broad-spectrum activity and can suppress growth of microbial competitors, or detoxify antiobiotics produced by some biocontrol microorganisms, as a self-defense mechanism against biocontrol agents.

Recently, it has been discovered that certain PGPB quench pathogen quorum-sensing capacity by degrading autoinducer signals, thereby blocking expression of numerous virulence genes. Since most, if not all, bacterial plant pathogens rely upon autoinducer-mediated quorum-sensing to turn on gene cascades for their key virulence factors (e.g. cell-degrading enzymes and phytotoxins), this approach holds tremendous potential for alleviating disease, even after the onset of infection, in a curative manner.

Although biocontrol activity of microorgansims involving synthesis of allelochemicals has been studied extensively with free-living rhizobacteria, similar mechanisms apply to endophytic bacteria, since they can also synthesize metabolites with antagonistic activity toward plant pathogens. For example, Castillo et al. demonstrated that munumbicins, antibiotics produced by the endophytic bacterium Streptomyces sp. strain NRRL 30562 isolated from Kennedia nigriscans, can inhibit in vitro growth of phytopathogenic fungi, P. ultimum, and F. oxysporum. Subsequently, it has been reported that certain endophytic bacteria isolated from field-grown potato plants can reduce the in vitro growth of Streptomyces scabies and Xanthomonas campestris through production of siderophore and antibiotic compounds. Interestingly, the ability to inhibit pathogen growth by endophytic bacteria, isolated from potato tubers, decreases as the bacteria colonize the host plant's interior, suggesting that bacterial adaptation to this habitat occurs within their host and may be tissue type and tissue site specific. Aino et al. have also reported that the endophytic P. fluorescens strain FPT 9601 can synthesize DAPG and deposit DAPG crystals around and in the roots of tomato, thus demonstrating that endophyte can produce antibiotics in planta.

Indirect Plant Growth Promotion through Induced Systemic Resistance

Biopriming plants with some PGPB can also provide systemic resistance against a broad spectrum of plant pathogens. Diseases of fungal, bacterial, and viral origin, and in some instances even damage caused by insects and nematodes, can be reduced after application of PGPB.

1. Induced systemic resistance: Certain bacteria trigger a phenomenon known as ISR phenotypically similar to systemic acquired resistance (SAR). SAR develops when plants successfully activate their defense mechanism in response to primary infection by a pathogen, notably when

the latter induces a hypersensitive reaction through which it becomes limited in a local necrotic lesion of brown, desiccated tissue. As SAR, ISR is effective against different types of pathogens but differs from SAR in that the inducing PGPB does not cause visible symptoms on the host plant. PGPB-elicited ISR was first observed on carnation (Dianthus caryophillus) with reduced susceptibility to wilt caused by Fusarium sp. and on cucumber (Cucumis sativus) with reduced susceptibility to foliar disease caused by Colletotrichum orbiculare. Manifestation of ISR is dependent on the combination of host plant and bacterial strain. Most reports of PGPB-mediated ISR involve free-living rhizobacterial strains, but endophytic bacteria have also been observed to have ISR activity. For example, ISR was triggered by P. fluorescens EP1 against red rot caused by Colletotrichum falcatum on sugarcane, Burkholderia phytofirmans PsJN against Botrytis cinerea on grapevine and Verticllium dahliae on tomato, P. denitrificans 1-15, P. putida 5-48 against Ceratocystis fagacearum on oak, P. fluorescens 63-28 against F. oxysporum f. sp. radicis-lycopersici on tomato and Pythium ultimum, F. oxysporum f. sp. pisi on pea roots, Bacillus pumilus SE34 against F. oxysporum f. sp. pisi on pea roots and F. oxysporum f. sp. vasinfectum on cotton roots.

2. Determinants of ISR: The ability to act as bioprotectants via ISR has been demonstrated for both rhizobacteria and bacterial endophytes, and considerable progress has been made in elucidating the mechanisms of plant-PGPB-pathogen interaction. Several bacterial traits (i.e., flagellation and production of siderophores and lipopolysaccharides) have been proposed to trigger ISR, but there is no compelling evidence for an overall ISR signal produced by bacteria. It has recently been reported that volatile organic compounds may play a key role in this process. For example, volatiles secreted by B. subtilis GBO3 and B. amyloquefaciens IN937a were able to activate an ISR pathway in Arabidopsis seedlings challenged with the soft-rot pathogen Erwinia carotovora subsp. carotovora. A major distinction often drawn between ISR and SAR is the dependence of the latter on the accumulation of salicylic acid (SA). Some PGPB do trigger an SA-dependent signaling pathway by producing nanogram amounts of SA in the rhizosphere. However, the majority of PGPB that activate ISR appear to do so via a SA-independent pathway involving jasmonate and ethylene signals. ISR is associated with an increase in sensitivity to these hormones rather than an increase in their production, which might lead to the activation of a partially different set of defense genes.

3. Defense mechanisms of ISR-mediated by PGPB: PGPB-triggered ISR fortifies plant cell wall strength and alters host physiology and metabolic responses, leading to an enhanced synthesis of plant defense chemicals upon challenge by pathogens and/or abiotic stress factors. After inoculation of tomato with endophytic P. fluorescens WCS417r, a thickening of the outer tangential and outermost part of the radial side of the first layer of cortical cell walls occurred when epidermal or hypodermal cells were colonized. In Burkholderia phytofirmans PsJN-grapevine interaction, a host defense reaction coinciding with phenolic compound accumulation and a strengthening of cell walls in the exodermis and in several cortical cell layers was also observed during endophytic colonization of the bacterium. The type of bacterized plant response induced after challenge with a pathogen resulted in the formation of structural barriers, such as thickened cell wall papillae due to the deposition of callose and the accumulation of phenolic compounds at the site of pathogen attack. Biochemical or physiological changes in plants include induced accumulation of pathogenesis-related proteins (PR proteins) such as PR-1, PR-2, chitinases, and some peroxidases. However, certain PGPB do not induce PR proteins but rather increase accumulation of peroxidase, phenylalanine ammonia lyase, phytoalexins, polyphenol oxidase, and/or chalcone synthase. Recent evidence

indicates that induction of some of these plant defense compounds (e.g. chalcone synthase) may be triggered by the same N-acyl homoserine lactones that bacteria use for intraspecific signaling. The revelation that some PGPB genes involved in antibiotic biosynthesis (e.g. phlD) are highly homologous with some plant genes involved in defense (e.g. chalcone synthase) raises the intriguing but as yet unexplored possibility that the products of these DeVriesien-like pangens may have interspecies activity benefiting plant protection, in addition to their currently known functions.

Biological Control of Decay Fungi in Wood

Forests cover about 3870 m ha of land, which accounts to 30% of earth. India with a forest area of 76.5 m ha is one of the largest producers of forest products, worth INR 30,000 crores per annum. Wood, the major forest product, is used for variety of purposes, such as building construction, composite manufacture, pulp and paper products, and a variety of finished materials including furniture, ports, pilings and distribution poles. It has been established that wood-rotting fungi, particularly basidiomycetes, damage forest wood even more than insects, marine animals or bacteria. These basidiomycetes are categorized as either white- or brown-rot fungi. Brown-rot fungi (BRF) decompose primarily the carbohydrate component of the wood. Their preferential decomposition of carbohydrate, cellulose and hemicellulose leads to fractures across the grains that break the wood into brown-coloured cubes. There is strong evidence to suggest that involvement of phenol-oxidizing enzymes during brown rot leads to generation of phenoxy radicals.

White-rot fungi (WRF), on the other hand, degrade both cellulose and lignin by secretion of cellulolytic and lignolytic enzymes. They decompose lignin of middle lamella sufficiently to cause separation of cells into fibers. The lignin is mineralized via extracellular fragmentation of the lignin polymer into lower molecular weight moieties that are then metabolized intracellularly. Over 600 species of WRF have been found to be lignolytic, converting lignin to carbon dioxide. Phanerochaete chrysosporium, a WRF, under nitrogen-limiting condition secretes at least six extracellular lignin peroxidase (LiP) and four manganese-dependent peroxidase (MnP) isozymes. Several other WRF secrete unique combinations of peroxidases and oxidases.

Trametes versicolor and Phlebia radiata produce one or more laccases in addition to LiP and MnP. Pleurotus sajor-caju secretes an aryl alcohol oxidase, a laccase and several peroxidases. Bjerkandera adusta secretes an aryl alcohol oxidase, while Rigidoporous lignosus and Dichomitus squalens (Polyporous anceps) secrete a laccase and MnP. MnP has also been implicated in the degradation of wood by the basidiomycete Neolentinus edodes. Until recently, the wood preserving industry has used three major preservatives, namely creosotes, pentachlorophenol and copper-chrome arsenate.

These are broad-spectrum preservatives but pose problems of environmental pollution from the disposal of toxic wastes and health hazard to workers. The immunological consequences of wood workers exposed to pentachlorophenol are reported to be activated T-cells, autoimmunity, functional immunosuppression and B-cell dysregulation. Besides adverse health and environmental effects, legislative constraints to these chemical preservatives suggest the need for new biological and biochemical control systems for preservation of wood.

Mechanism of Biocontrol

Antibiosis and mycoparasitism, the two direct mechanisms of antagonism, have been proposed to explain the inhibition of fungal pathogens in the rhizosphere by biocontrol agents. Antibiosis occurs when one or more diffusible compounds inhibit growth or developmental changes in the pathogen, impairing its ability to colonize the rhizosphere and establish disease. Mycoparasitism, on the other hand, is a different process initiated by physical destruction of fungal cell wall mediated by the action of hydrolytic enzymes produced by the biocontrol agent. A third mechanism, involving competition for space and nutrients within the rhizosphere, has also been of significance.

Evidences for the role of competition and parasitism in the biocontrol of plant diseases have been convincing. However, antibiosis is much less clearly established due to the lack of methods for a meaningful evaluation of the production and function of antibiotics in soil. Fungal cell wall degrading enzymes produced by an antagonist are, therefore, thought to be involved simultaneously in parasitism and antibiosis. Antibiosis, nevertheless, is an advantageous mechanism particularly for biological control of diseases, because compounds mediating antibiosis can diffuse rapidly in nature and direct contact between the antagonist and pathogen is not necessary.

Many antibiotics produced by actinomycetes have been used directly, or assumed to be responsible, for the biocontrol activity of the producing strain. Such metabolites may include macrolide, benzoquinones, aminoglycosides, polyenes and nucleoside antibiotics. Certain species of actinomycetes, such as Streptomyces, are known to produce not only antibiotics but also extracellular enzymes active in fungal cell wall degradation, such as β-1,3-glucanases and chitinases. The role of these enzymes in antifungal activity and biocontrol is of interest to several research groups. Recently, the authors have shown that Streptomyces violaceusniger strain XL-2 exhibits strong antagonism towards white- and brown-rot fungi, Postia placenta, Phanerochaete chrysosporium, Trametes versicolor and Gloeophyllum trabeum. The antagonism is ascribed to the extracellular production of a protein, which is yet to be characterized, though it has the potential to be exploited commercially in healthcare and agriculture industry.

During the last decade, chitinases have received increased attention because of their wide range of applications. Chitin is found in the cuticle of insects and the shells of crustaceans and mollusks. Also, the cell walls of most taxonomic groups of fungi contain chitin. Since the fungal cell walls are rich in chitin, the potential application of chitinase in biocontrol of fungal phytopathogens is promising. Chitinase producing microbial strains might be used directly in biological control of fungi, or indirectly by using purified protein or through gene manipulation.

Tree-derived Phenolic Compounds as Control Agents for Wood-decaying Basidiomycetes.

Naturally occurring phenolic compounds were used as possible regulators of fungal growth. Taylor et al reported that the growth of Trametes versicolor on wood was affected in a bimodal fashion via timedependent application of catechol. Data from other studies also indicated that phenolic compounds (quinones) might regulate hyphal growth in a bimodal fashion through the products of extracellular polyphenol oxidase.

The effect of 12 monomeric aromatic compounds on the production of six carbohydrate-degrading enzymes from two BRF, Postia (Oligoporous) placenta and Gloeophyllum trabeum, and one WRF,

Trametes versicolor was reported by Highley and Micales. Most compounds at a concentration of 0.05% (w/v) were inhibitory to the growth of the decay fungi. When incorporated into the liquid growth medium of the fungi, some of these compounds inhibited the production of enzymes. Catechol and vanillin (50 ppm) caused complete inhibition of xylanase and β-1,4-endoglucanase production by P. placenta. No aromatic monomer, however, strongly inhibited all enzyme activities of all of the fungi. Interestingly, the efficacy of phenolic fraction of the Hopea parviflora heartwood and Cashew nut shell liquid was investigated against termites and fungi by Sharma et al. They found the results encouraging; however, further examination for the efficacy of phenolic compounds as fungal growth control agents is required.

Microfungi as Biocontrol Agents

Much research has centred on the use of Trichoderma species as potential control agents for a wide range of plant pathogens. Trichoderma spp. are known for rapid colonization of wood, production of lytic enzymes including chitinase and β-1,3- glucanase, and the secretion of volatile and nonvolatile fungicidal and fungistatic secondary metabolites.

A commercially available mixture of Trichoderma polysporum and T. harzianum (Binab) was used by Morrell and Sexton to control decay of loblolly pine sapwood and Douglas-fir heartwood by five basidiomycete fungi. Although they achieved good biocontrol against Neolentinus lepideus and other BRF when wood was out of ground contact, most of the test fungi were not completely inhibited. The biocontrol agent had little effect on WRF, or even on N. lepideus, when wood was exposed to soil. According to the authors, Binab did not appear to be a feasible method for controlling decay of Douglas fir or southern pine poles.

Bruce and Highley tested fifteen isolates of Trichoderma, two of Penicillum, and one of Aspergillus as biocontrol agents against two basidiomycetes, Trametes versicolor and Neolentinus (Lentinus) lepideus. Although the filtrates for most isolates produced some growth inhibition, only three of the isolates (two Trichoderma and one Aspergillus) produced sufficient effects against sixteen brown- and white-rot basidiomycetes. The culture filtrates of all three isolates showed varying degrees of inhibition against each of the basidiomycetes; filtrates of the Aspergillus producing the best inhibition against WRF. In another study, Bruce et al reported that distribution pole interiors colonized by Trichoderma were able to resist decay by N. lepideus and Antrodia carbonica, the two most common basidiomycetes attack pine and Douglas-fir distribution poles. Although there was a reduction in the level of decay produced by T. versicolor over a twelve-week incubation period in a glass jar experiment, the Trichoderma treated poles were found susceptible to decay by T. versicolor in nature. These results and the earlier findings by Highley and Richard clearly show that wood blocks pretreated with Trichoderma (Binab) were resistant to decay by BRF but offered little protection against WRF.

Bacteria as Biocontrol Agents

Bacteria, though major contributors to the decay of waterlogged wood, are much less important than fungi as agents of wood degradation and as such do not represent a significant target for biological control systems. Over the past fifteen years, however, promising results have emerged by using bacteria to control sapwood-inhabiting blue-stain fungi. Benko and Highley evaluated the effectiveness of the bacterial cultures against blue stain and mold fungi as well as BRF and WRF.

They used a mixed bacterial solution consisting of six bacteria from the genera Pseudomonas (P. cepacia), Streptomyces (S. chrestomyceticus, S. rimosus and S. rimosus forma paromomycinus), Streptoverticillium (S. cinnamoneum forma azacoluta), and Xenorhabdus (X. luminescens). The mixed bacterial culture was found strongly antagonistic against the wood-attacking fungi. Southern yellow pine (Pinus spp.) blocks treated with the solution of mixed bacterial culture suffered less than 1% weight loss after two months' exposure to the BRF (Postia placenta) or WRF (T. versicolor). Laboratory tests also indicated complete inhibition of blue-stain fungus, Ceratocystis coerulescens or mold, Trichoderma harzianum over the same period of time.

Actinomycetes as Biocontrol Agents

Over the past 55 years, actinomycetes have been the most widely exploited group of microorganisms in the production of secondary metabolites of commercial importance in medical and agricultural applications. Actinomycetes and particularly Streptomyces spp. are good sources of novel antibiotics, enzymes, enzyme inhibitors, immunomodifiers and vitamins. Their ubiquitous nature and prolific metabolic activity has led to 4,607 patents for actinomycetes-related products, including 3,477 antibiotics produced from Streptomyces alone. Actinomycetes are Gram-positive, filamentous bacteria that are among the most abundant soil and rhizosphere microorganisms. Like filamentous fungi they grow with branching hyphae and can penetrate insoluble substrates, such as lignocellulose. Some of the examples of common genera of lignocellulose-degrading actinomycetes are Streptomyces, Micromonospora, Microbispora, Thormomonospora, Norcardia and Arthrobacter spp. Lignin degradation is a primary metabolic activity in the case of Streptomyces in contrast to Phanerochaete chrysosporium, where it is a secondary metabolic activity.

Streptomyces are important saprophytic soil microorganisms and well-known producers of antibiotics and extracellular enzymes. They are primarily degraders of grass-type lignocelluloses. Streptomyces spp. solubilize lignin but their mineralization of lignin to CO_2 is much less than that of other WRF. This low wood lignin mineralization ability of Streptomyces spp. means that Streptomyces and other actinomycetes may be useful as biocontrol agents without much concern over their wood-decaying ability. Their biocontrol abilities clearly correlate with the production of antibiotics. Streptomyces violaceusniger YCED9, for example, is a soil isolate which exhibits biocontrol activity against a variety of plant pathogenic fungi. The strain produces at least three antifungal antibiotics, including Nigericin, Geldanamycin and a complex of polyenes that includes Guanidylfungin . Streptomyces spp. are also known for their ability to cause lysis of fungal hyphae by producing chitinases and glucanases as already mentioned. The antifungal biocontrol agent, S. lydicus WYECwas capable of not only destroying germinating oospores of Pythium ultimum but also damaging the cell walls of the fungal hyphae. WYEC108 also produced high levels of chitinases, induced to high levels as fungal cell walls are used as a carbon source in growth media. However, negligible levels of enzymes were detected when S. lydicus WYEC108 was grown in the absence of chitin. Chitinase production by S. lydicus WYEC108 was also induced by colloidal chitin, N-acetylglucosamine and chito-oligosaccharides. However, the synthesis was repressed by high (but not low) levels of glucose and carboxy methyl cellulose (CMC).

Actinomycetes Fb was reported to possess antagonistic activity against fungi, Aureobasidium pullulans and Hormonea dematodes. Many such reports are available in the literature that is related to the production of antibiotics antagonistic to several other fungi. These antibiotic substances induce malformations in fungi, such as stunting, distortion, swelling, hyphal protuberances or the highly branched appearance of fungal germ tubes, an indirect evidence to show antibiosis as a mechanism of antagonism. Using such criteria, it was detected that antibiotics of some soil actinomycetes caused similar effects on hyphae of Helminthosporium sativum, in culture and in soil. Several species from Streptomyces violaceusniger clade produced antifungal antibiotics, such as Niphithricin, Spirofungin, Azalomycin F complex, Guanidylfungins and Malonylniphimycin.

Soil and aquatic actinomycetes show considerable ability to survive starvation. Antibiotics and protein inhibitors are formed during the late growth cycle, when familiar regulatory processes, like transcriptional control, are ineffective. These secondary metabolites can prevent degradation of enzymes and structuralproteins essential for survival as well as biosynthesis, which might form aberrant products during nutrient limitation. Interestingly, secondary metabolite production in Streptomyces spp. is subject to catabolic repression in the presence of high levels of carbon and nitrogen sources. Repression of secondary metabolite biosynthesis by ammonia or certain amino acids is common in actinomycetes. Nitrogen limiting conditions lead to the secretion of ligninases responsible for wood degradation by WRF, like P. chrysosporium. Low nitrogen conditions would also be conducive to the secretion of antifungal and antibacterial secondary metabolites by actinomycetes used as biocontrol agents.

References

- Plant-disease, science: britannica.com, Retrieved 2 May, 2019

- An-introduction-to-plant-pathogen: creative-diagnostics.com, Retrieved 7 January, 2019

- Food-supply: carleton.edu, Retrieved 14 August, 2019

- Main-types-of-penetration-of-post-harvest-pathogens-plant-diseases, plant-diseases, plants: biologydiscussion.com, Retrieved 22 April, 2019

- Pathogen: bionet.nsc.ru, Retrieved 25 February, 2019

- Diseases-caused-by-virus-in-plants-with-control-measures, plants: biologydiscussion.com, Retrieved 27 July, 2019

- List-of-diseases-in-plants-caused-by-viruses: hunker.com, Retrieved 7 March, 2019

- BiologicalControl: apsnet.org, Retrieved 11 June, 2019

- Nematodes, pathogens: entomology.cornell.edu, Retrieved 10 March, 2019

3

Biological Control of Fruits Diseases

There are various fungal and bacterial diseases which plague fruit bearing trees and plants. A few examples of such diseases are fire blight, peach scab and brown rot. The topics elaborated in this chapter will help in gaining a better perspective about the diseases which affect different fruit crops as well as how they can be treated using biological control.

Fruit Diseases

Pests and diseases of fruit trees abound. During the first summer, that fruit-trees have enemies and that they don't need just cultivation and feeding, but also protection. In another page we have pictures of some of the garden pests being described below for identification purposes.

Pests and Diseases of Fruit Trees: Aphids

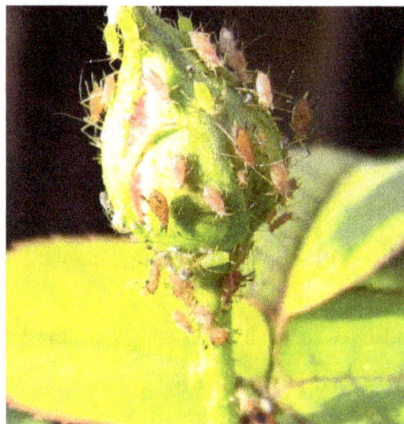

The black and green aphids, or plant-lice, are often very troublesome.

Aphids appear in immense numbers on the young and tender shoots of trees, and by sucking their juices check or enfeeble the growth. They are the milch-cows of ants, which are usually found very busy among them.

Nature apparently has made ample provision for this pest, for it has been estimated that "one aphid individual in five generations might be the progenitor of six thousand millions." Aphids are easily destroyed. Prepare a barrel of tobacco juice by steeping stems for several days, until the juice is of a dark brown color; we then mix this with soap-suds.

A pail is filled, and the ends of the shoots, where the insects are assembled, are bent down and dipped in the liquid. One dip is enough. Such parts as cannot be dipped are sprinkled liberally with a garden-syringe, and the application repeated from time to time, as long as any of the aphids remain.

The liquid can be so strong that it can damage the leaves; therefore it is better to test it on one or two subjects before using it extensively. Apply it in the evening.

Pests and Diseases of Fruit Trees: Apple Scale

Apple scale attacks weak, feeble-growing trees, and can usually be removed by scrubbing the bark with the preparation given above.

Pests and Diseases of Fruit Trees: Codling Moth

The codling moth, or apple a codling moth worm is another enemy that should be fought resolutely, for it destroys millions of pounds of fruit. Who has not seen the ground covered with premature and decaying fruit in July, August, and September? Each specimen will be found perforated by a worm hole. The egg has been laid in the calyx of the young apple, where it soon hatches into a small white grub, which burrows into the core, throwing out behind it a brownish powder.

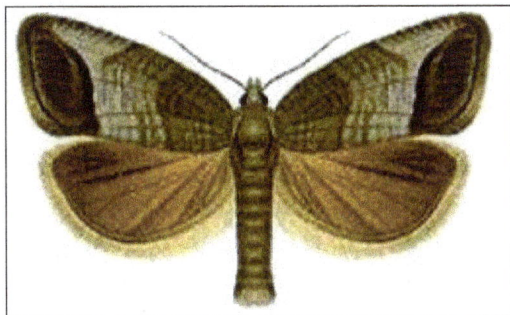

After about three weeks of eating the apple it eats its way out, shelters itself under the scaly bark of the tree. If allowed to be scaly, or in some other hiding-place, spins a cocoon, and in about three weeks comes out a moth, and is ready to help destroy other apples. This insect probably constitutes one of nature's methods of preventing trees from overbearing; but it so exaggerates its mission that it has become an insufferable nuisance.

Natural control of codling moth recommend that trees should be scraped free of all scales in the spring, and washed with a solution of soft soap. About the 1st of July, wrap bandages of old cloth, carpet, or rags of any kind around the trunk and larger limbs.

The apple worms will appreciate such excellent cover, and will swarm into these hiding places to undergo transformation into moths. Therefore the wraps of rags should be taken down often, thrown into scalding water, dried, and replaced. The fruit as it falls should be picked up at once and carried to the pigs, and, when practicable, worm-infested specimens should be taken from the trees before the worm escapes.

Pests and Diseases of Fruit Trees: Canker Worm

The canker worm in those localities where it is destructive can be guarded against by bands of tar-covered canvas around the trees. The moth cannot fly, but crawls up the tree in the late autumn and during mild spells in winter, but especially throughout the spring until May.

Pests and Diseases of Fruit Trees: Tent Caterpillar

We have all seen the flaunting, unsightly abodes of the tent caterpillar and the foliage-denuded branches about them. Fortunately these are not stealthy enemies. You only to look very early in the morning or late in the evening to find them all bunched up in their nests. These should be taken down and destroyed.

Pests and Diseases of Fruit Trees: Cherry and Pear Slugs

Cherry and pear slugs, "small, slimy, dark brown worms," can be destroyed by dusting the trees with dry wood ashes or air-slacked lime.

Pests and Diseases of Fruit Trees: Mice

Field mice are often found around young trees, especially during the winter, working beneath the snow. Unless heaps of rubbish are left here and there as shelter for these little pests, one or two good cats will keep the acre free of them. Treading the snow compactly around the tree should also be done from time to time.

Pests and Diseases of Fruit Trees: Apple Scale

Finally be nice to the birds. They are the best of all insect destroyers, along with frogs in a pond tht should be close-by, or having your orchard enclosed and allowing a family of Khaki ducks to feed on fallen fruit and insects. Put up plenty of bird houses for backyard birds such as bluebirds and wrens, and treat the little brown song-sparrow as one of your stanchest friends.

Peach Scab

Peach scab caused by Cladosporium carpophilum is a common stone fruit pathogen, and is primarily a problem in warm, humid areas of production. For New York growers, this means that peach scab is not likely to severely affect peach plantings. Unfortunately, the problem has become more severe since winters have been warmer than usual in the last decade. Overall, peach scab is often the most damaging on peach and tart cherry, but is sometimes found on plums and apricots. Late bearing varieties are often more severely infected than earlier varieties.

Peach Scab is often found near the stem end of the fruit

Symptoms and Signs

Twigs and leaves can be infected by C. carpophilum, but the lesions are rarely noticeable. The lesions are on the undersides of infected tissues and are often the same color as the tissue. Olive green lesions are often very discernable on half formed fruit, and are typically found closer to the stem. These lesions blacken with age, and may coalesce as they enlarge. Severe infections crack the fruit.

Fruit infection varies by the Prunus species affected. Nectarine fruit lesions are often considerably larger than any other examples, and sour cherry fruit lesions are often red instead of olive green.

Impact and Considerations

Severe infections crack the fruit open and allow the entry of secondary rots.

Twig and leaf infections don't do enough damage to be noticed, and don't change color. In warmer climate orchards, these symptoms are often so numerous that sanitation is unfeasible. Fruit lesions can become a much more significant problem once they coalesce and crack the fruit. This opens the fruit to brown rot, which can subsequently infect all surrounding fruit. Fruit infections also have a lengthy incubation (40-70 days), meaning that lesion development can occur before the infection becomes apparent.

Epidemiological Aspects

- C. carpophilum overwinters as mycelia in twig lesions and as chlamydospores on bark.

- Overwintering lesions are sporulating two weeks before calyx ('shuck') split. The number of conidia drastically increases in the four weeks following calyx split.

- Most conidia are produced during humid weather, and this period is the most conducive for infection.

Peach scab is often found near the stem end of the fruit.

Peach Leaf Curl

Peach leaf curl disease is distributed throughout the world wherever peach is grown. The disease received its name on account of its characteristic effect on leaf —curl. There is no definite knowledge of the place of origin of the disease. But it is believed to have been introduced in the European countries far back in 1841 from China, the original home of the peach.

The disease is now widely distributed in Europe and in the United States, in parts of Asia, China and Japan, arid also in Africa Australia, and New Zealand.

Symptoms of Peach Leaf Curl

The disease affects leaves, tender growing shoots and very rarely flower parts and fruits. It is most conspicuous on the leaves.

The disease first appears although relatively inconspicuous in symptoms, in the early spring when the leaves begin to unfold. Soon after the leaves are well out of the bud some of them appear to be distorted, puckered along the midrib and curled up assuming a red tint. Leaf blade and petiole may both be involved in the curling.

While in more heavily infected leaves the curling is so severe that the whole lamina with the exception of the tip looks like a partially inflated paper bag. The blistered portions of the leaf are softer than the disease-free parts of the leaf blade. Only a few leaves of a tree may be affected or the infections may be so numerous as to involve almost the entire foliage.

The individual lesions may include a small portion of a leaf or its entire surface. As the disease progresses, the leaf tissue becomes yellowish and shows some reddish colour. The reddish-velvety surface of the lamina soon becomes whitish-grey. The leaves gradually turn brown, wither and fall off.

In heavy infections the trees may suffer severely from premature defoliation. The loss of leaves is recovered by a fresh crop of leaves produced from dormant buds. Twigs may also be affected. Infected twigs become pale-green to yellow, swollen, stunted, sometimes exude gummy material and ultimately die. Flowers and fruits when infected drop promptly.

Causal Organism of Peach Leaf Curl

Taphrina deformans (Berk.) Tul. That the disease is caused by an ascomycete, at that time called Exoascus deformans was first described by the Rev. M. J. Barkeley. The old generic name Exoascus is still in use by many authors. The fungus does not produce any ascocarp.

An intercellular, septate mycelium is quite common in the infected host tissue. The intercellular mycelium travels across the mesophyll and proceeds to spread out chiefly between the palisade layer and the upper epidermis, where it develops in greater abundance than in any others parts of the leaf.

In preparation for the reproductive stage, the branches of the mycelium which become established between the epidermis and mesophyll how make their way between the epidermal cells to form a network of short cells beneath the cuticle.

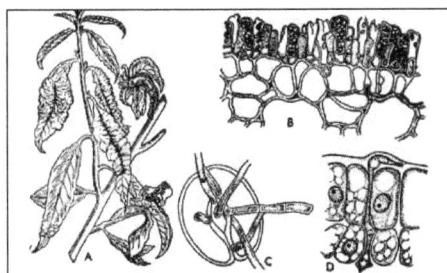

Peach leaf curl. A. Twing with diseased leaves. B. Section through diseased leaf showing arrangement of asci. C. Host penetration. D. Hyphal ramification in the host tissue.

The cells of the hyphae are with dense cytoplasm and two nuclei. The subcutaneous and intercellular hyphal cells become uninucleate resulting from karyogamy. These are the ascogenous cells from which asci are developed. The upward growth of the developing asci raises the cuticle, and this is either pierced or torn and disappears, leaving the asci exposed to the surface as a more or less continuous hymenium.

The asci appear as a palisade-like layer. The asci are usually flattened or somewhat truncate at the free end, broader above than below usually eight unicellular ascopores are formed in each ascus, but occasionally the ascospore number may be only two.

The ascospores are forcibly expelled through an apical slit or rupture in the ascus and may accumulate on the surface of the leaf, giving a white or greyish powdery condition.

They commonly germinate by pudding to form a series of uninucleate thin-walled secondary spores which are also known as conidia. The ascospores may also germinate directly by the formation of term tubes. The young hyphae become binucleate in the host tissue.

Disease Cycle of Peach Leaf Curl

The incidence of the disease is associated with comparatively low temperatures such as occur in the spring. The optimum temperature for the growth of the fungus is about 20 °C. Opinions differ with regard to the method by which the fungus survives from season to season.

For a long time it was believed that the fungus existed as a perennial mycelium in the twigs, the fungus attacking the leaves again at the opening of the buds. The general opinion is that the primary infections which occur at the opening of the buds in spring are effected by spores, probably by conidia, which have been sheltered on some part of the host or may also persist in the soil or in the soil cover.

Germ tubes from ascospores or conidia penetrate the young leaf directly into the cuticle or pass through the stoma. The fungus makes its initial growth within the cuticular layer and then sends hyphae between the epidermal cells. Ultimately the hyphae become established as intercellular mycelium.

The first effect of the mycelium is to irritate the host cells and stimulate them to an abnormal activity. The infected host cells increase in size and change in form and structure. The affected cells suffer almost a complete loss of chlorophyll accompanied with puckering with the midrib behaving like a puckering string.

Peach being a host of economic importance, economic losses at one time may be very heavy. But this is one of the diseases which is rather easily controlled with the adoption of proper protective measures.

Diseases cycle of Peach leaf curl is presented in figure below.

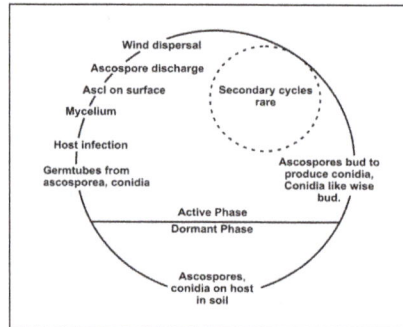

Disease cycle of peach leaf curl.

Brown Rot

Brown rot is the most important disease risk for stone fruits in warm, humid climates. It is the primary disease for which fungicides are applied to stone fruits.

Brown rot of stone fruits

PATHOGEN: Monilinia fructicola, M. laxa, and M. fructigena.

HOSTS: Stone fruits (peach, nectarine, cherry, plum), almond, and occasionally some pome fruits (apple and pear)

Symptoms and Signs

The brown rot fungi cause a blight of blossoms and twigs figure and a soft decay of fruits of peaches, cherries, and plums. Thus, there are two distinct phases of this disease.

Blossom and Twig Blight

This phase of the disease occurs in early spring when the trees are blooming, although twig blight also can occur during the fruit rot phase. The anthers and pistil of the flower are infected initially. The fungus then invades the floral tube, the ovary, peduncle, and usually the twig to which the peduncle is attached. Infected blossoms wilt, turn brown, and usually cling to the twig. Extension of the infection into the peduncle and twig results in a necrotic area in the woody tissue termed a "canker". Sometimes succulent twigs and shoots become infected directly when there are extended periods of both moisture and warm temperatures (20 to 28 °C, 68 to 82 °F).

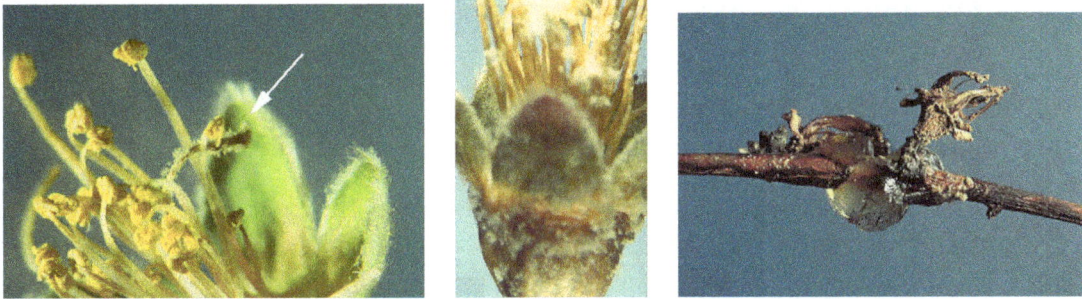

Under moist or humid conditions, ash-gray-brown colored sporodochia (tufts of conidiophores) bearing conidia (asexual spores) form on the surface of diseased blossoms and twigs. The presence of conidia is a diagnostic sign that separates brown rot from other fungal and bacterial diseases of stone fruits. A gummy substance usually exudes from the cankers, causing the blighted flowers to adhere to the twig.

Fruit Rot

Fruit susceptibility to brown rot increases during the 2 to 3 week period prior to harvest. Increased susceptibility is associated with an increase in sugar content as the fruits ripen. Initially, tan-brown, circular spots are visible on the fruit. Under humid conditions, ash-gray-brown masses of conidia develop on these lesions. There can be thousands of conidia on a lesion, each potentially capable of initiating a new infection. If environmental conditions are wet and warm during fruit ripening, the entire crop can literally be destroyed"overnight".

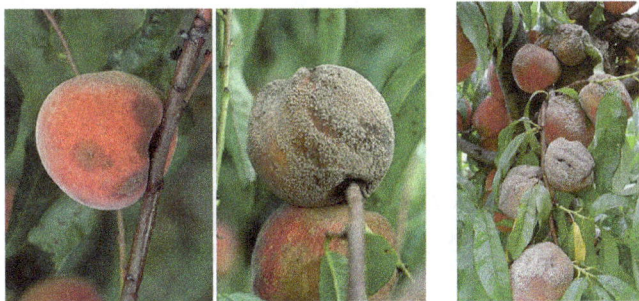

Diseased fruit that do not fall to the ground dehydrate and become shriveled "mummies" that cling to the branch. Sometimes the fungal infection extends from the fruit into the twig and branch. Although not common, brown rot also can occur on ripe apples and pears.

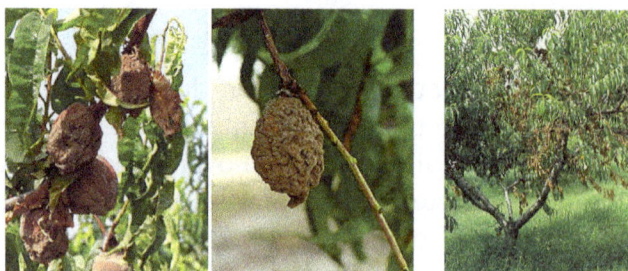

Pathogen Biology

Monilinia Fructicola, M. laxa and M. Fructigena

The first published description of a brown rot fungus on decaying fruit was in 1796. Different species of brown rot fungi were later discovered. The fruit-rotting fungi were placed in the genus Sclerotinia in the late 1800s and were transferred to the new genus Monilinia in 1928.

Of the three closely related fungal species causing brown rot, M. fructicola is the species most commonly found in North America, Australia, New Zealand, Japan, Brazil and other South American countries. Monilinia laxa is found in all major fruit producing countries where brown rot occurs. Monilinia laxa is relatively widespread in California and also occurs in the midwestern and northeastern states, but has not been found in the southeastern states. This species is especially common in Europe, South Africa, and Chile. The disease it causes is frequently called "European brown rot." Monilinia fructigena occurs on both stone and pome fruits in Europe, but does not cause the extensive crop loss caused by M. laxa and M. fructicola. The mycelial growth characteristics of the three Monilinia species vary when they are grown on 2% potato-dextrose agar (PDA) medium. Isolates of M. fructicola can have different fruit rotting and sporulation capabilities.

Asexual reproduction

Conidia (asexual spores) are produced on tufts of conidiophores called sporodochia. The conidia are hyaline (colorless), lemon-shaped, and produced in a moniloid manner (resembling a string of beads with constricted ends). Under ideal conditions, conidia germinate within 3 to 5 hours. Extensive mycelial growth can occur within 24 hours.

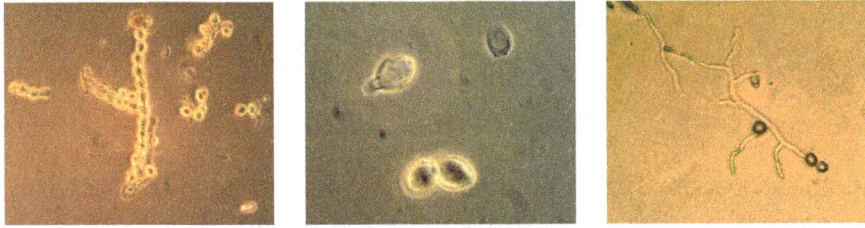

Sexual Reproduction

Brown rot fungi are ascomycetes. They produce ascospores (sexual spores) in tubular sacs termed asci that are produced on the upper surface of a cup- or disc-shaped structure, known as an apothecium. Apothecia can be 5-20 mm (up to nearly an inch) in diameter and are borne on mummified fruit that have fallen to the ground. Although commonly described in textbooks, apothecia are rarely observed in most areas.

Disease Cycle and Epidemiology

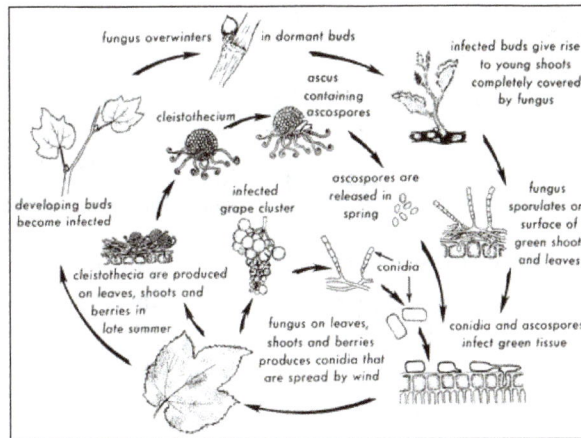

Disease Cycle

Epidemiology

The opening blossoms are the first emerging susceptible tissue in the spring. Sources of blossom blight inoculum are mummies, infected peduncles, and cankers. Conidia from these sources are disseminated by splashing or wind-blown rain.

Infrequently, apothecia develop from mummies on the orchard floor or beneath secondary hosts such as wild plums and other Prunus spp. surrounding the orchard. Ascospores are released during rainfall and are carried by wind to the blossoms.

Blossom infection is highly dependent on wetness duration and temperature. For blossom infection to occur at 10 °C (50 °F), 18 hours of wetting are necessary; in contrast, at 24 °C (77 °F) only 5 hours are necessary. The time required for symptoms of blossom blight to develop may be only a few days to 1 or 2 weeks depending on the temperature. Blighted blossoms often are obscured as new flushes of leaf growth occur. Thus, the first evidence that infection has occurred may be yellow and wilting leaves on branches or twigs. It takes very few blighted blossoms to cause severe fruit rot if environmental conditions are optimal as fruit ripen.

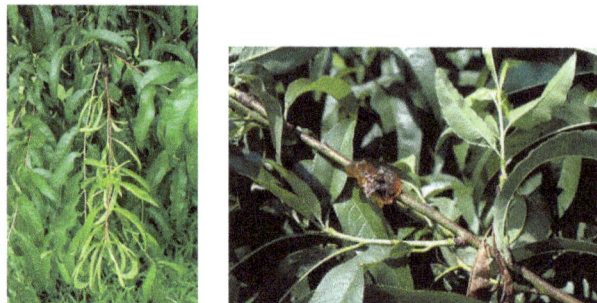

As fruit ripen and the sugar content increases, they become increasingly susceptible to infection. In contrast, green, immature fruit are less prone to infection unless they are injured. Inoculum sources for fruit infection include blighted blossoms, cankers, mummies from the previous year, and diseased fruit in the tree or on the orchard floor from thinning practices. The amount of inoculum is very important in determining the severity of brown rot. Warm, wet or humid weather during the 2 to 3 week period prior to harvest increases disease severity because it increases both the level of inoculum and the amount of infection. If wet weather extends into the harvest period, fruit loss can be severe. Insects, such as June beetles, which are attracted to overripe fruit, can increase disease severity not only by carrying the fungal conidia but also by creating wounds as they feed. Wet, warm conditions also increase overwintering inoculum available for blossom infections the following spring.

Fire Blight

Fire blight is a common and very destructive bacterial disease of apples and pears. The disease is caused by the bacterium Erwinia amylovora, which can infect and cause severe damage to many plants in the rose (Rosaceae) family. On apples and pears, the disease can kill blossoms, fruit, shoots, twigs, branches and entire trees. While young trees can be killed in a single season, older trees can survive several years, even with continuous dieback.

Fire blight damage on an apple tree.

Table List of commonly grown plants in Ohio that are susceptible to fire blight	
Apple	Mountain Ash
Blackberry	Pyracantha
Cotoneaster	Quince
Crabapple	Raspberry
Hawthorn	Spirea

Disease Development

Fire blight first appears in the spring when temperatures get above 65 degrees F. Rain, heavy dews and high humidity favor infection. Precise environmental conditions are needed for infection to occur and as a result disease incidence varies considerably from year to year.

Fire blight bacteria overwinter as cankers in living tissue on the trunk and main branches and on mummified fruit. Primary infections are initiated during bloom when bacteria are carried from the cankers to open flowers by splashing rain, pollinating insects (i.e., bees, pollen wasps, flies, ants) or during production practices such as pruning. Relatively few overwintering cankers become active and produce bacteria in the spring, but a single active canker may produce millions of bacteria, enough to infect an entire orchard. These bacteria multiply rapidly in the blossom nectar, and spread to the spurs (blossom-bearing twigs), new shoots and branches, resulting in secondary infections. Shoot infections can also occur through wounds created by sucking insects (aphids, leafhoppers or tarnished plant bugs), freeze or frost damage, wind whipping, wind-driven rain, or hail. Once a shoot is infected, the fire blight bacteria multiply rapidly and droplets of ooze can be seen within three days. Shoots remain highly susceptible to infection until vegetative growth ceases and the terminal bud is formed.

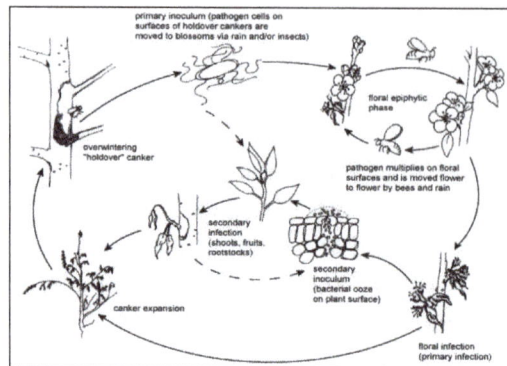

Fire blight disease cycle.

Symptoms

Fire blight symptoms vary depending on the tissue affected and can also vary between pear and apple.

Blossom and Spur Blight

Blossom and spur symptoms appear in the spring. Bacteria gain entry into the tree via blossoms and new shoots. Diseased blossoms become water-soaked, wilt and turn brown. Bacteria spread rapidly into other flowers in the cluster and then move down into the spur. Spurs become blighted, turning brown on apples and black on pear.

Fire blight on apple blossoms and spurs.

Shoot Blight

Shoot blight starts at the growing tips of shoots and moves rapidly down into older portions of the twig. Blighted twigs first appear water-soaked, then turn dark brown or black. As blighted shoots wilt, the twigs bend at the growing point and resemble a shepherd's crook or an upside down "J".

Blighted leaves remain attached to the dead branches throughout the summer. During warm and humid weather infected shoots will ooze droplets of creamy white bacteria.

Fire blight on an apple twig. Curved "shepherd's crook"
at the tip of the diseased twigs.

Stem Cankers

As fire blight bacteria move through blighted twigs into the main branches, the bark sometimes cracks along the margin of the infected area causing a distinct canker. Bark on younger trees becomes water soaked and the cankers have a dark brown to purple color. Sapwood beneath a canker has a reddish brown appearance and may be soft to the touch. Cankers can girdle the main branches and trunk causing additional dieback.

Fire blight cankers on young apple tree trunks (top) and mature tree branch (left).
Reddish brown appearance of the sapwood beneath the canker on the mature tree branch.

Fruit Blight

Both apple and pear fruit may be blighted. Rotted areas turn brown to black and become covered with droplets of whitish tan colored bacterial ooze. Fruit remain firm and eventually dry out and shrivel into mummies.

Rootstock Symptoms

Fire blight symptoms on rootstocks usually develop near the graft union. Symptoms are similar to those of stem cankers. Fire blight infections in rootstocks can rapidly kill the tree by girdling the rootstock.

Postharvest Diseases in Fruit Crops

Postharvest disease may occur at any time during postharvest handling, from harvest to consumption. Losses caused by postharvest diseases are greater than generally realized because the value of fresh fruits increases several fold while passing from the field to the consumer. Postharvest losses are estimated to range from 10 to 30% per year despite the use of modern storage facilities and techniques. When estimating postharvest disease losses, it is important to consider reductions in fruit quantity and quality, as some diseases may not render produce unsalable yet still reduce product value. Postharvest diseases affect a wide variety of crops particularly in developing countries which lack sophisticated postharvest storage facilities. Infection by fungi and bacteria may occur during the growing season, at harvest time, during handling, storage, transport and marketing, or even after purchase by the consumer. It is also important to take into account costs such as harvesting, packaging and transport when determining the value of produce lost as a result of postharvest wastage. Aside from direct economic considerations, diseased produce poses a potential health risk. A number of fungal genera such as Penicillium, Alternaria and Fusarium are known to produce mycotoxins under certain conditions. Generally speaking, the greatest risk of mycotoxin contamination occurs when diseased produce is used in the production of processed food or animal feed.

Losses due to postharvest disease are affected by a great number of factors including:

1.   Commodity type.

2.   Cultivar susceptibility to postharvest disease.

3.   Postharvest environment (temperature, relative humidity, atmosphere composition, etc).

4.   Produce maturity and ripeness stage.

5.   Treatments used for disease control.

6.   Produce handling methods.

7.   Postharvest hygiene.

Virtually all postharvest diseases of fruit are caused by fungi and bacteria. In general, however, viruses are not an important cause of postharvest disease. The so-called 'quiescent' or 'latent' infections are those where the pathogen initiates infection of the host at some point in time (usually before harvest), but then enters a period of inactivity or dormancy until the physiological status of the host tissue changes in such a way that infection can proceed. Examples of postharvest diseases arising from quiescent infections include anthracnose of various tropical fruit caused by Colletotrichum spp. and grey mould of strawberry caused by Botrytis cinerea. The other major groups of postharvest diseases are those which arise from infections initiated during and after harvest.

Postharvest diseases of fruit crops: Fruit crops are attacked by a wide range of microorganisms in the postharvest phase. Actual disease only occurs when the attacking pathogen starts to actively grow in the host. Diseases are loosely classified according to their signs and symptoms. Signs are visible growths of the causal agents, and symptoms the discernible responses produced by the host. Postharvest diseases are caused primarily by microscopic bacteria and fungi, with fungi the most important causal agent in fruit crops. Fungi are further subdivided into classes and are described as lower fungi, characterized by the production of sporangia which give rise to numerous sporangiospores, or higher fungi, described as ascomycetes, deuteromycetes, and basidiomycetes.

Ascomycetes are exemplified by fruiting bodies that release sexual spores when mature. Deuteromycetes, a form of ascomycetes, only release asexual spores. They are more common than the sexual ascomycetes stage in postharvest crops. Deuteromycetes are further subdivided into hyphomycetes and coelomycetes based on spore and structural characteristics. The agonomycetes contain important soil pathogens that form survival structures known as sclerotia that allow them to survive in the absence of the host lists many important diseases of fruit crops according to host and causal agents.

Causes of postharvest disease of fruit crops: Correct identification of the pathogen causing postharvest disease is central to the selection of an appropriate disease control strategy. Many of the fungi which cause postharvest disease belong to the phylum Ascomycota and the associated fungi Anamorphici. In the case of the Ascomycota, the asexual stage of fungus (anamorph) is usually encountered more frequently in postharvest diseases than the sexual stage of the fungus (teleomorph). Important genera of anamorphic postharvest pathogens include Penicillium, Aspergillus, Geotrichum, Botrytis, Fusarium, Alternaria, Colletotrichum, Dothiorella, Lasiodiplodia and Phomopsis.

In the phylum Oomycota, the genera Phytophthora and Pythium are important postharvest pathogens, causing a number of diseases such as brown rot in citrus (Phytophthora citrophthora and P. parasitica). Rhizopus and Mucor are important genera of postharvest pathogens in the phylum Zygornycota. R. stolonifer is a common wound pathogen of a very wide range of fruit, causing a rapidly spreading watery soft rot. Genera within the phylum Basidiomycota are generally not important causal agents of postharvest disease, although fungi such as Sclerotium rolfsii and Rhizoctonia solani. The major causal agents of bacterial soft rots are various species of Erwinia, Pseudomonas, Bacillus, Lactobacillus and Xanthomonas. Bacterial soft rots are very important postharvest diseases generally of less importance in most fruit.

Host physiological status: The development of postharvest disease is intimately associated with the physiological status of the host tissue. To create the right environment for minimizing postharvest losses due to disease it is important to understand the physiological changes that occur after produce is harvested. All plant organs undergo the physiological processes of growth, development and senescence. Growth and development generally only occur while the organ is attached to the plant (with the exception of seed germination and sprouting of storage organs), but senescence will occur regardless of whether the organ is attached or not. When an organ such as a fruit is harvested from a plant, it continues to respire and transpire depleting both food reserves and water. Maturity is a term often used in reference to fruit and is frequently confused with the term ripeness.

Fruit are often classified into two groups on the basis of how they ripen: Climacteric fruit exhibit a pronounced increase in respiration and ethylene production coincidentally with ripening. Climacteric fruit can be harvested in an unripe state and providing they are sufficiently mature, will ripen to an acceptable quality. Non-climacteric fruit do not exhibit a rapid increase in respiration during the ripening process. The eating quality of non-climacteric fruit does not improve after harvest, although they may undergo some changes in colour development and softening. For this reason they should not be harvested until they are common postharvest diseases and pathogens of fruit crops.

1. Temperate fruits

    - Pome fruit

        - Disease- Gray mould, Bitter rot, Alternaria rot

        - Pathogen - Botrytis cinerea, Colletrotrichum gloeosporioides, Alternaria spp.

    - Stone fruit

        - Disease- Brown rot, Grey mould, Blue mould, Alternaria rot

        - Pathogen- Monilia spp., Botrytis cinerea, Penicillium spp., Alternaria alternate.

    - Grapes

        - Disease- Grey mould, Blue mould

        - Pathogen- Botrytis cinerea, Penicillium spp.

    - Berries

        Disease- Grey mould, Cladosporium rot, Blue mould

        Pathogen- Botrytis cinerea, Cladosporium spp. Penicillium spp.

2. Subtropical fruit

    - Citrus fruit

        - Disease- Green mould, Black centre rot Stem end rot

        - Pathogen- Penicillium digitatum, Alternaria citri, Phomopsis citri.

    - Avocado

        - Disease- Anthracnose, Anthracnose, Stem end rot, Bacterial soft rot

        - Pathogen- Colletotrichum gloeosporioides, Colletotrichum acutatum, Dothiorella spp., Erwinia carotovora.

3. Tropical fruit

    - Banana

        - Disease- Crown rot, Black end, Ceratocystis fruit rot

        - Pathogen- Fusarium spp., Nigrospora spharica, Thielaviopsis paradoxa.

    - Mango

        - Disease- Anthracnose, Stem end rot, Black mould, Alternaria rot, Grey mould, Blue mould

        - Pathogen- Colletrotrichum gloeosporioides, Phomopsis mangifera, Aspergillus niger, Alternaria alternate, Botrytis cinerea, Penicillium expansum.

- Pawpaw (Papaya)

  - Disease- Anthracnose, Black rot, Phomopsis rot

  - Pathogen- Colletrotrichum spp., Phoma caricae papaya, Phomopsis caricae papaya.

- Pine apple

  - Disease- Water blister, Fruit let core rot, Yeasty rot, Bacterial brown rot

  - Pathogen- Thielaviopsis paradoxa, Penicillium funiculosum, Saccharomyces spp., Erwinia ananas.

Table: Classification of some common edible fruit is according to respiratory behavior during ripening.

S. No.	Climacteric fruit	Non-climacteric fruit
1	Apple	Blackberry
2	Apricot	Carambola
3	Avocado	Cherry
4	Banana	Grape
5	Blueberry	Grapefruit
6	Custard apple	Lemon
7	Guava	Lime
8	Kiwi fruit	Longan
9	Mango	Litchi
10	Pawpaw (Papaya)	Mandarin
11	Passion-fruit	Orange
12	Peach	Pineapple
13	Pear	Raspberry
14	Plum	strawberry

Mode of infection: Infection of fruit by postharvest pathogens can occur before, during or after harvest. Infections which occur before harvest and then remain quiescent until some point during ripening are particularly common amongst tropical fruit crops. Anthracnose, which is the most serious postharvest disease of a wide range of tropical and sub-tropical fruit such as mango, banana, papaya and avocado, is an example of a disease arising from quiescent infections established prior to harvest. Various species of Colletotrichum can cause anthracnose. In avocado for example, early studies reported that ungerminated appressoria were the quiescent phase of C. gloeosporioides. Studies conducted two decades later however showed that appressoria germinated to produce infection hyphae prior to the onset of quiescence. In any case, the fungus ceases growth soon after appressorium formation and remains in a quiescent state until fruit ripening occurs.

In avocados, antifungal dienes are present in the peel of unripe fruit at concentrations inhibitory to Colletotrichum gloeosporioides, the avocado anthracnose pathogen. Grey mould of strawberry caused by Botrytis cinerea is another important postharvest disease sometimes arising from quiescent infections established before harvest. Conidia of B. cinerea on the surface of necrotic flower parts germinate in the presence of moisture. The fungus colonises the necrotic tissue and then remains quiescent in the base of the floral receptacle. Many postharvest diseases develop from the stem end of fruit. The mode of infection involved in this group of diseases can however vary considerably. In the example of B. cinerea, lesions occurring at the stem end of fruit arise from quiescent floral infections. Stem end rots of citrus caused by Lasiodiplodia theobromae and Phomopsiscitri result from quiescent infections in the stem button of fruit. In other stem end rot diseases, infection occurs during and after harvest through the wound created by severing the fruit from the plant (e.g. banana crown rot). Endophytic infection, where by the fungus symptomlessly and systemically colonises the stem, inflorescence and fruit pedicel tissue, is important in a number of stem end rot diseases of tropical fruit. Mango stem end rot caused by Dothiorella dominicana is one example of a postharvest disease arising from endophytic colonisation of fruit pedicel tissue. In this case, the fungus colonises the pedicel and stem end tissue of unripe fruit, where it remains quiescent until fruit ripening commences.

Postharvest diseases which can arise from late season infections include brown rot of peach (Monilinia fructicola) and grey mould of grape (Botrytis cinerea). Mechanical injuries such as cuts, abrasions, pressure damage and impact damage commonly occur during harvesting and handling. Some chemical treatments used after harvest, such as fumigants used in insect disinfestations and disinfectants such as chlorine, may also injure produce if applied incorrectly. Tropical fruit in particular are very sensitive to low temperatures, many developing symptoms of chilling injury below 13 °C (depending on storage duration). For example, the incidence of alternaria rot in pawpaw and apple is increased by exposure to excessive cold. For example, hot water dipping of mangoes for excessive times or temperatures can result in increased levels of stem end rot (Dothioretla spp.).

Preharvest Factors that Influence Postharvest Diseases

1. Weather: Weather affects many factors related to plant diseases, from the amount of inoculum that overwinters successfully to the amount of pesticide residue that remains on the crop at harvest. Abundant inoculum and favorable conditions for infection during the season often result in heavy infection by the time the produce is harvested. For example, conidia of the fungus that causes bull's-eye rot are rain dispersed from cankers and infected bark to fruit especially if rainfall is prolonged near harvest time, causing rotten fruit in cold storage several months later.

2. Physiological condition: Condition of produce at harvest determines how long the crop can be safely stored. For example, apples are picked slightly immature to ensure that they can be stored safely for several months. The onset of ripening and senescence in various fruits renders them more susceptible to infection by pathogens. On the other hand, fruits can be made less prone to decay by management of crop nutrition.

3. Fungicide Sprays: Certain pre-harvest sprays are known to reduce decay in storage. Several studies have been done on the effectiveness of pre-harvest ziram fungicide application on pome fruit and show an average reduction in decay of about 25 to 50% with a single spray. Iprodione has been used for several years as a pre-harvest spray 1 day before harvest to prevent infection of stone fruit by Monilinia spp. In combination with wax and/or oil its decay control spectrum is increased and it will also control postharvest fungi such as Rhizopus, and Alternaria. The new class of strobilurin fungicides promises to provide postharvest control of several diseases in fruits. They are especially effective against fruit scab on apples and should reduce the presence of pin point scab in storage.

Postharvest Factors that Influence Decay

1. Packing Sanitation: It is important to maintain sanitary conditions in all areas where produce is packed. Organic matter (culls, extraneous plant parts, soil) can act as substrates for decay-causing pathogens. For example, in apple and pear packinghouses, the flumes and dump tank accumulate spores and may act as sources of contamination if steps are not taken to destroy or remove them. Chlorine readily kills microorganisms suspended I dump tanks and flumes if the amount of available chlorine is adequate. A level of 50 to 100 ppm of active chlorine provides excellent fungicidal activity. Chlorine measured as hypochlorous acid can be obtained by adding chlorine gas, sodium hypochlorite, or dry calcium hypochlorite. Although chlorine effectively kills spores in water it does not protect wounded tissue against subsequent infection from spores lodged in wounds. Organic matter in the water inactivates chlorine, and levels of chlorine must be constantly monitored. Recently, in precisely controlled tests in water or as foam, chlorine dioxide was found to be effective against common postharvest decay fungi on fruit packinghouse surfaces.

2. Postharvest Treatments: Products used for postharvest decay control should only be used after the following critical points are considered:

Type of pathogen involved in the decay:

1. Location of the pathogen in the produce.

2. Best time for application of the treatment.

3. Maturity of the host.

4. Environment during storage, transportation and marketing of produce. Specific materials are selected based on these conditions and fall into either chemical or biological categories listed below.

3. Fungicide treatments: Several fungicides are presently used as postharvest treatments for control of a wide spectrum of decay-causing microorganisms. However, when compared to preharvest pest control products the number is very small. For example, intensive and continuous use of fungicides for control of blue and green mold on citrus has led to resistance by the causal pathogens of these diseases. Chemical treatments that are presently used are thiabendazole, dichloran, and imazalil. However, resistance to thiabendazole and imazalil is widespread and their use as effective materials is declining. These products include sodium benzoate, the parabens, sorbic acid, propionic acid, SO_2, acetic acid, nitrites and nitrates, and antibiotics such as nisin.

4. Biological control: Postharvest biological control is a relatively new approach and offers several advantages over conventional biological control:

1. Exact environmental conditions can be established and maintained.

2. The bio-control agent can be targeted much more efficiently.

3. Expensive control procedures are cost-effective on harvested food.

Several biological control agents have been developed in recent years, and a few have actually been registered for use on fruit crops. The first biological control agent developed for postharvest use was a strain of Bacillus subtitles. It controlled peach brown rot, but when a commercial formulation of the bacterium was made, adequate disease control was not obtained.

5. Irradiation: Although ultraviolet light has a lethal effect on bacteria and fungi that are exposed to the direct rays, there is no evidence that it reduces decay of packaged fruits. More recently, low doses of ultraviolet light irradiation (254 nm UV-C) reduced postharvest brown rot of peaches. In this case, the low dose ultraviolet light treatments had two effects on brown rot development; reduction in the inoculum of the pathogen and induced resistance in the host. Gamma radiation has been studied for controlling decay, disinfestation, and extending the storage and shelf-life of fresh fruits. A dose of 250 Gy has an adverse effect on grapefruits increasing skin pitting, scald, and decay. Low doses of 150 for fruit flies and 250 gray (Gy) for codling moth are acceptable quarantine procedures. Gamma irradiation may be used more in the future once methyl bromide is no longer available to control insect infestation in stored products. All uses of methyl bromide are being phased out to avoid any further damage to the protective layer of ozone surrounding the earth.

6. Temperature and relative humidity: Proper management of temperature is so critical to post-harvest disease control that all other treatments can be considered as supplements to refrigeration. Fruit rot fungi generally grow optimally at 20 to 25 °C (68 to 77 °F) and can be conveniently divided into those with a growth minimum of 5 to 100 °C (41 to 50 °F) or -6 to 0 °C (21.2 to 32 °F). Fungi with a minimum growth temperature below -20C (28.40F) cannot be completely stopped by refrigeration without freezing fruit (Shine, et al. 2007). High temperature may be used to control postharvest decay on crops that are injured by low temperatures such as mango and papaya. Heating of pears at temperatures from 21 to 38 °C (69.8 to 100.4 °F) for 1 to 7 days reduced postharvest decay. Decay in "Golden Delicious" apples was reduced by exposure to 38 °C (100.4 °F) for 4 days and virtually eliminated when treated after inoculation.

7. Modified or controlled atmospheres: Alterations in O_2 and CO_2 concentrations are sometimes provided around fruit and vegetables. Because the pathogen respires as does produce, lowering the O_2 or raising the CO_2 above 5% can suppress pathogenic growth in the host. In crops such as stone fruits, a direct suppression occurs when fungal respiration and growth are reduced by the high CO_2 of the modified atmosphere. Low O_2 does not appreciably suppress fungal growth until the concentration is below 2%. Important growth reductions result if the O_2 is lowered to 1% or lower although there is a danger that the crop will start respiring an aerobically and develop off-flavor.

Integrated Control of Postharvest Diseases

Effective and consistent control of storage diseases is dependent upon integration of the following practices:

1. Select disease resistant cultivars where possible.
2. Maintain correct crop nutrition by use of leaf and soil analysis.
3. Irrigate based on crop requirements and avoid overhead irrigation.
4. Apply pre-harvest treatments to control insects and diseases.
5. Harvest the crop at the correct maturity for storage.
6. Apply postharvest treatments to disinfest and control diseases and disorders on produce.
7. Maintain good sanitation in packing areas and keep dump water free of contamination.

8. Store produce under conditions least conducive to growth of pathogens. Integration of cultural methods and biological treatments with yeast biocontrols has been studied on pears. These results demonstrated that unrelated cultural and biological methods that influenced pear decay susceptibility can be combined into an integrated program to substantially reduce decay. As a general rule, alternatives to chemical control are often less effective than many fungicides. Therefore, it will generally be necessary to combine several alternative methods to develop an integrated strategy to successfully reduce postharvest decay.

Biological Control in Orchards

Green lacewing eggs are laid on the tips of long, white, hair-like stalks to prevent cannibalism. The larvae (called aphid lions) are generalist predators of mites, thrips, soft scales, and almost any other soft-bodied prey.

It has been difficult to utilize the full potential of biological control in tree fruit and other crops that receive periodic sprays of broad-spectrum pesticides and/or have high quality standards. The best pest targets for biological control in tree fruits are generally the secondary foliage-feeding pests that do not cause direct fruit injury (e.g. mites, aphids, and leafminers). Populations of pests that feed directly on the fruit (e.g. codling moth, Oriental fruit moth, and plum curculio) generally cannot be tolerated at levels high enough for special biological control agents to reproduce.

Natural enemies and environmental factors limit populations of insect and mite pests in natural ecosystems. When natural enemies are killed by human's actions in any habitat or when pests are introduced to new habitats without their natural enemies, natural control often fails and results in pest outbreaks.

While biological control is often thought of as a biopesticide where a single species of beneficial arthropod is released or conserved, the best results are most often achieved where a complex of many species of natural enemies, including predators and parasitoids, each contribute to reducing pest populations at different times of the season and on different developmental stages. While the development of pesticide resistance (mainly to organophosphates) has occurred in Stethorus punctum, the black ladybeetle predator, and several species of predatory mites, such resistance is generally much slower to develop in beneficial arthropods. The biological control potential of the vast majority beneficial arthropods is not realized unless they develop resistance to pesticides, no pesticides are used, or only pesticides that are selective and nontoxic to these arthropods are used.

Complex of natural enemies for woolly apple aphid.

Types of Biological Control Agents

Predators

- Consume many prey during development.

- Generally larger than prey.

- All stages may be predators.

- Are often generalists rather than specialists on any one prey type and eat both adults and immatures.

Parasitoids

- Immatures feed only on a single host and almost always kill it.

- Are smaller than the host.

- Are often specialized in their choice of host species and life stages thereof.

- Only the female attacks the host and lays eggs or larvae on or in the host.

- Immatures remain on or in the host, adults are free living and mobile and may be predaceous, feed on nectar, or not feed at all.

Parasites

- Smaller than host and don't generally kill it (e.g. mites).

Pathogens

- Diseases caused by fungi, bacteria, and viruses that kill the host.

- Some are naturally occurring and some have been commercially developed.

- Bacillus thuriengensis (Bt) toxins and spores--Dipel, etc.

- Fermentation products from fungi are precursors to making abamenctin and spinosad.

- Codling moth polyhedrosis virus available commercially for control.

- Naturally occurring Beauvaria and Hirsutella fungal pathogens.

Biological Control of Mites

The most successful biological control programs in eastern tree fruits have centered on the conservation of native species of mite predators to control the European red mite and twospotted spider mite. After 40 years of use, some of these predators have developed resistance to organophosphate insecticides (e.g. Stethorus), but are suppressed or eliminated when broad-spectrum carbamate and pyrethroid insecticides are used. The use of pheromone mating disruption, horticultural oils, and some of the more selective reduced-risk insecticides and miticides will allow a natural increase of predators capable of regulating pest mite populations to tolerable levels without the use of miticides. Mite control through biological control has the additional advantage of stopping the development of miticide resistance and, once established, is sustainable long-term if the use of certain harmful pesticides is avoided. The routine use of carbamates and pyrethroids in stone fruits, pears, grapes, and small fruits currently prevents reliable biological mite control agents even though many of the same predators found in apples can be present.

The main biological mite predators found in Pennsylvania apple orchards are listed below:

Typhlodromus pyri (Phytoseiidae)

Discovered in Pennsylvania for the first time in 2003, this predatory mite is currently the most reliable and effective mite predator in eastern U.S. apple orchards. Pear shaped and slightly larger than a European red mite adult, they are white/translucent until they feed. When feeding on adult red mites or apple rust mites, its abdomen may appear reddish. T. pyri is very active and moves rapidly to consume up to 350 mite prey in a lifespan of about 75 days. Females may lay up to 70 eggs each and have several generations per season. Populations, therefore, can build rapidly in response to pest mite populations. Most effective in the cooler weather of the spring and fall, T. pyri is somewhat less effective in the summer months. It overwinters on the apple tree under the bark where it is less susceptible to dormant oil applications and is very tolerant of Pennsylvania's winters.

T. pyri for biological control of European red mite.

Neoseiulus Fallacis (Phytoseiidae)

Almost indistinguishable from T. pyri except under a microscope, this predator is currently more widespread in distribution in Pennsylvania apple orchards than T. pyri, due to a higher tolerance for some pesticides and the use of alternative plant hosts. Like T. pyri, N. fallacis is also very active, but is able to build populations three times faster during the hotter summer months. This predator lives only about 20 days with each female laying 40 to 60 eggs and may have 6 to 7 generations/year. Like T. pyri, N. fallacis is resistant to organophosphate insecticides, but it is very susceptible to pyrethroids and carbamates. This predator is not as tolerant of cool weather in the spring and fall and is susceptible to winter kill in Pennsylvania. Purely a predator, N. fallacis is not able to co-exist on apple trees without pest spider mite populations to consume and will often leave the tree to feed on mites in the orchard ground cover.

Zetzellia Mali (Stigmaeidae)

An omnivore like T. pyri that is able to exist on pollen, fungi, and rust mites when spider mite populations are absent, Z. mali is very slow moving and feeds only on the eggs of pest and predatory mites. Its diamond shape and bright yellow coloration (turning more reddish after feeding) make it easy to distinguish this predator from other predatory mites. It is smaller in size than the European red mite. Because it is less active, it is able to exist on pest mite populations even lower than T. pyri. Like T. pyri, it is also more active in the cooler spring and fall months. However, with only a couple of generations each season and a consumption rate of only two to three eggs per day, it cannot usually be relied on to control mite pests alone. It is a valuable supplement to control by other mite predators and is much more tolerant of most pesticides, including carbamates and pyrethroids. Generally, populations of more than one per leaf are necessary to exert significant control of spider mite populations.

Stethorus Punctum (Coccinellidae)

Once the cornerstone of biological mite control in Pennsylvania apple orchards, this small, black ladybeetle predator has greatly declined in importance over the last five years. Although one of the smallest of all ladybird beetles, S. punctum was the most important beneficial insect in Pennsylvania apple orchards starting in the mid-1970s and conservation of this predator reduced miticide use by 50 percent for over 30 years. While tolerant of many organophosphate insecticides, the decline of this predator was mainly due to the greater use of pyrethroids and the introduction of several new neonicotinoid and IGR insecticides that are toxic to various life stages of this predator. Reproducing only when populations of pest mites exceed eight to ten mites per leaf, relying on S. punctum alone requires grower tolerance of some foliar mite injury. With the registration of newer, more effective miticides in recent years, most growers are not willing to tolerate this injury, despite the high cost of miticides. S. punctum is now much less common in orchards and generally in small localized "hot spots" of mites. The main advantage of this predator is its ability to fly and quickly colonize areas of high mite populations.

Biological Control of Aphids

Aphid Midge Aphidoletes aphidimza (Cecidomyiidea)

The aphid midge, Aphidoletes aphidimyza, often contributes to biological control of spirea and green aphids in pome fruits. They feed on many species of aphids on many type of crops, but are not generally found in stone fruits because of their susceptibility to pyrethroids. Generally tolerant to organophosphate insecticides as immatures and slightly less so as adults, all stages are susceptible to carbamates, pyrethroids, neonicotinoid, and certain miticides. This species can be reared and is sold from biological control companies for mass releases in many crops, but especially for aphid control in greenhouses.

Ladybird Beetles (Coccinellidae)

Adults from these easily recognized beetles are oval, often brightly colored and spotted, and vary in size from 1.5 mm to 6 mm. Approximately a dozen of the 450 species found in North America are found in fruit with most feeding primarily on aphids, but some like Stethorus specialize on mites while others specialize on scales and mealybugs. A number of species require pollen as adults to reproduce and some can be important predators of moth eggs.

Multicolored Asian Ladybird Beetles-Harmonia Axyridis

The multicolored Asian ladybird beetle has recently become the most common and most effective aphid predator in Pennsylvania orchards, replacing Coccinella septumpunctata and several native species. H. axyridis is native to Asia, but was released in Pennsylvania in 1978 and 1981. However, overwintering individuals were not recorded until 1993, and the populations that have become established may have resulted from an accidental introduction by an Asian freighter in New Orleans.

Multi-colored Asian Ladybird beetle.

Green and Brown Lacewings (Chrysopidae and Haemorobiidae)

Green lacewing adults are 6/10 to 9/10 inch in length, green with transparent wings with an interconnecting network of fine veins. The many different species are difficult to distinguish, but the adult of the most common green lacewing species has golden eyes. The adults feed on nectar, honeydew, and pollen with females producing 400 to 500 eggs each over a relatively long life of up to 3 months. Green lacewing eggs are laid on the tips of long, white, hair-like stalks to prevent cannibalism. The larvae (called aphid lions) are generalist predators of mites, thrips, soft scales, and almost any other soft-bodied prey. They are voracious aphid predators, eating 100 to 600 aphids during a 1 to 2 week development period and can be important predators of moth eggs and larvae as well. Prey are seized in hollow, sickle-like jaws protruding from the head and sucked dry. The larvae make a small, round, and white pupal case, often on the stem or calyx end of the fruit where they overwinter or, in the case of one species, overwinter as adults in bark crevices and other protected places. Brown lacewings are smaller (1/5 to 6/10 inch long) and are predatory, both as adults and larvae. They are much more tolerant of colder weather than the green lacewings and are more useful predators early in the season. Females lay 100 to 460 eggs, but not on stalks like the green lacewings. Larvae may consume more than 20 aphids per day or 30 to 40 mites per day. Developmental times are slower with most species only having two generations per season. Both types of lacewings have some tolerance to organophosphate insecticides, but should be conserved by selective pesticide use.

Minute Pirate Bug-Orious insidiosus

Generalist predators of aphids and mites, these are very small 1/10 inch, black, somewhat oval-shaped bugs that look like miniature, dark, tarnished plant bugs. They are most easily recognized by white, shiny wing patches on the adults. Able to feed on a wide variety of small prey, including thrips, leafhoppers, moth eggs, and young larvae, they are able to subsist on pollen or plant juices when prey are not available. This habit of feeding on plant juices may make them more susceptible to plant systemic products like some neonicotinoid insecticides. They are efficient at searching out high-prey densities and will aggregate where there is an abundance of prey. When handled, Pirate Bugs are capable of causing a mild sting with their beak. Orius has several generations/year and take about 20 days to develop from egg to adult. The adults live about 35 days with each female inserting about 130 eggs into plant tissues. Immature stages and adults can eat about 30 mites/aphids per day. Adults appear in late April, continue to feed all season until early fall, and then overwinter in the leaf litter both inside and outside orchards. They have some tolerance to organophosphate insecticides, but should be conserved by selective pesticide use.

Syrphid Flies

Several species of syrphid flies are among the most voracious of aphid predators in Pennsylvania orchards. Adults are known as hover flies and resemble bees except that they have only one pair of wings. They are generally brown to black with yellowish areas. Their food source is pollen, nectar, and aphid honeydew, which is necessary for proper development of the eggs. Eggs are white, elliptical, and less than 4/100 inch long. The larvae, or maggots, are elongate, tapering gradually toward the head end and may be cream, yellow, gray, or a combination of these colors. Adults lay eggs in the midst of aphid colonies. Larvae cast their head side to side to locate aphids, which they pierce and consume. A single larva may destroy hundreds of aphids as it completes its three development stages in about 3 weeks. There may be five to seven generations per year with most species overwintering as adults or last instar larvae. Check for the presence of eggs and larvae in aphid colonies. Control of green aphids may result if 20 percent of the aphid colonies have syrphid larvae present.

Sand Wasps Crabronidae

Several species of these wasps are specialist predators of stink bugs and actively queue in on their smell to find their host. Bycertes quadrifasciata is a species that looks like a smaller version of the cicada killer wasp whose nests in the sand have been found to have prey consisting almost entirely of brown marmorated stink bug nymphs from which the wasps develop quite normally. Each female is capable of collecting up to 50 nymphs, which they paralyze and place in several burrows 6-8 inches into sandy ground. Another species, Astata unicolor, has also been seen to collect nymphs, but nests have not yet been found. This species is commonly found around our orchards as it prefers to nest in heavier ground. Like most predatory and parasitic wasps, it needs to visit flowers as adults to obtain nectar for food to mature its eggs.

Lepidopteran Predators-Ground Beetles

(Carabidae) and Rove Beetles (Staphylinidae)

These are two of the largest families of beetles with 1,500 ground beetle and 3,000 rove beetle species in North America. Many are generalist predators that are effective in controlling pests that pupate or overwinter in the ground cover or on the trunks (e.g. codling moth, Oriental fruit moth, apple maggot, plum curculio, European apple sawfly, leafrollers). Many live in the ground cover away from pesticide applications made to tree foliage, but some may climb trunks. All are very pesticide susceptible and are often used as indicators of environmental quality.

Woolly Apple Aphid Parasitoid-Aphelinus Mali

These adult wasps are very small and they insert their eggs singly into the body of aphids, where they will develop internally to kill the host. There are six to seven generations each year with each generation taking about 20 to 25 days to develop. Larvae or pupae overwinter within the mummified body of the aphid. A. mali are most effective in reducing small woolly apple aphid colonies in the spring when colonies are small. If biological control is disrupted with toxic pesticides, A. mali are less effective in controlling larger colonies later in the season. These very small wasps are able to attack only the aphids on the periphery of the colony and cannot successfully penetrate the wax and mass of aphid bodies to attack the center of the aphid colony, thus the percentage of parasitism actually decreases as the aphid colonies get larger in size. From midsummer to late season woolly apple aphid colonies are usually brought under control by a complex of syrphid fly species and generalist predators, such as brown and green lacewings. Ladybug larvae and adults are occasional predators of woolly apple aphids but do not appear able to deal with the waxy covering and give little control.

Lepidopteran and Stink Bug Parasitoids

Tachinid Flies: Important parasitoids of leafrollers in the spring. One species, Actia interrupta, is currently the most important parasitoid of the obliquebanded leafroller. Eggs are laid on the skin of larvae to hatch and develop externally on the larvae to eventually leave just an empty husk of skin. Pupae are generally found near the host remains and resemble a grain of wheat in size and shape. Trichopoda pennipes, an important parsitoid of the squash bug, is also now undergoing a host shift to attack the brown marmorated stink bug with eggs being present at levels of up to 20 percent in some locations, but successful development on this host remains low so far. All species appear to be very susceptible to pesticides and are important only in pheromone disruption or orchards with minimal pesticide sprays.

Scelionid Wasps: Several species of Trissolcus and Telenomus wasps are egg parasitoids of our native stink bug species, are the primary regulatory agents of these pests outside our orchards, and reduce their numbers so they are of minor importance when they move into the orchards from other hosts. Several species are in the process of undergoing host shifts to the brown marmorated stink bug, but current rates of parasitism arc under 5 percent. A specialist Trissolcus parasitoid of BMSB in Asia, where it parasitizes up to 80 percent of eggs, is being evaluated by USDA-ARS for possible mass release sometime in the future.

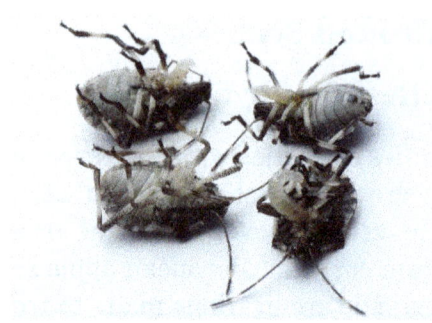

Egg parasitoids of stink bug species.

Braconid and Ichneumon Wasps: With approximately 120,000 known species and many as yet undescribed, this is a virtually untapped source of biological control in modern agriculture. With various complex life histories, often alternating between several hosts and attacking specific life stages, these wasps have not been important sources of biological control in tree fruit since the introduction of disruptive broad-spectrum insecticides. Previous to this, however, they provided almost complete control of many of the leafroller species. Currently, there are more than 40 different wasp parasitoids capable of attacking tufted apple bud moth in Pennsylvania apple orchards. All species appear to be very susceptible to pesticides and are important only in pheromone disruption or orchards with minimal pesticide sprays. Braconid species appear to be most important late in the growing season.

Trichogramma Egg Parasitoid: Most commonly employed as a biopesticides obtained from biological supply houses for mass releases into many crops. These tiny wasps complete their development inside a single egg of their moth or butterfly host. Native populations of mostly T. minutum attack many different orchard pests in Pennsylvania (most important are the several species of leafrollers, codling moth, and oriental fruit moth). The life of the adults and the number of eggs laid are greatly increased with the provision of nectar sources and females may then live up to 2 weeks and lay over 80 eggs. Although present during most of the growing season, populations generally do not build to be significantly important in controlling these orchard pests until late summer. Trichogramma is very susceptible to pesticides and is important only in pheromone disruption or orchards with minimal pesticide sprays.

Fungicide Resistance Management

Resistance has sometimes resulted in pest-management-program failures. Below are presented tactics to help delay resistance to fungicides.

Understanding Types of Fungicides

Pesticides used for managing fungi-caused fruit diseases are either fungicidal (they kill fungi) or fungistatic (they inhibit fungal growth). Fungicides can be separated into two categories: protectants and systemics.

Protectant fungicides protect the plant against infection at the site of application. Their characteristics are as follows:

- They provide protection against infection.

- They do not penetrate into the plant.

- They require uniform distribution over the plant surface.

- They require repeated application to renew deposit.

- They have a multisite mode of action against fungi.

- Fungi are not likely to become resistant to protectant fungicides. Some common protectant fungicides are Bravo, captan, copper, Dithane, Manzate, Polyram, sulfur, and Ziram.

Systemic fungicides prevent disease from developing on parts of the plant away from the site of application. Their characteristics are as follows:

- They penetrate into the plant.

- They move within the plant.

- They control disease by protectant and/or curative action.

- They often have a very specific mode of action against fungi. Some systemic fungicides are Elite, Flint, Indar, Rally, Merivon, Orbit, Pristine, Procure, Rubigan, and Sovran.

Cultural Control and Fungicide use Patterns

Due to environmental conditions, disease is inevitable in the Mid-Atlantic growing region and use of chemical controls is a necessity; however, following cultural practices that favor decreasing disease pressure will help decrease the opportunity for resistance. Using resistant varieties, minimizing tree stress, and maintaining proper soil fertility reduces disease incidence since pathogens do not reproduce well on trees that are less susceptible to disease. As a result, the chance of resistance decreases. Avoid selecting sites with high disease pressure since this increases the chance of selecting for resistant fungi. Using dormant copper sprays and removing inoculum sources such as leaves (using urea or a flail mower), mummified fruit, and dead twigs/branches reduces the initial pathogen population. When using fungicides, use only when needed since this avoids unnecessary selection for resistant populations. It is important to be sure sprayers are appropriately calibrated and covering trees effectively. Achieving good spray coverage, tank-mixing with protectants, and alternating fungicides with different modes of action (FRAC group) reduces populations exposed to selection.

Fungicide Resistance Issues and Mitigation Strategies for Specific Diseases

Apple Scab and Brown Rot

Fungicides in FRAC Groups 3, 7, 9, and 11 are highly effective against scab infection on apples and brown rot on stone fruit. However, apple scab and brown rot fungi can become resistant to these fungicides, especially if any of them are continually applied alone. Growers using one of these fungicides to control apple scab or brown rot must be certain to not only alternate it with an unrelated fungicide but also use it in combination with a broad-spectrum fungicide, like captan, metiram (Polyram), mancozeb, Ziram, thiram, sulfur, or ferbam. Another strategy to prevent resistance is to alternate the use of these materials throughout the season. The less any one of them is used in an orchard during a given season, the lower the chances that resistance will develop. At the present time, we know fungi causing apple scab and brown rot have shown high tolerance to fungicides in FRAC Group 11. As a result, growers are cautioned when using fungicides in this class, especially when it is not included as a premix. These fungicides should be avoided during peak primary apple scab spore dispersal, which is from late pink through petal fall.

Mitigating Fungicide Resistance for Apple Scab

Using cultural controls, such as removing inoculum sources (fallen leaves), is important for decreasing disease incidence; however, during seasons where the disease pressure is high (frequent rains, warm temperatures), fungicide applications will be important. It is critical to monitor disease conditions since this will play a crucial role in deciding which fungicides to use and when. In addition, if the alternate row middle (ARM) method is being used, it is very important to not stretch intervals, especially during frequent warm and rainy conditions. Sometimes this may mean shrinking intervals to five days, especially if disease conditions are favorable. There have been incidences where apple scab "broke through" as a result of stretching ARM intervals too long during very wet periods.

From Green Tip through Tight Cluster

Scab spores will begin to be dispersed from overwintering leaves starting at green tip; however, the spore numbers will be low, gradually increasing over time. If conditions are dry, focus on managing powdery mildew by using products such as Indar, Rally, Topguard/Rhyme, or sulfur tank-mixed with a broad-spectrum fungicide (EBDC, ferbam, metiram, ziram). Dry weather plus low scab spore numbers equals low disease pressure. Although some strong powdery mildew products are not as effective against scab, a broad-spectrum fungicide will keep the disease in check. If disease conditions are favorable for scab (warm and wet), then consider using other fungicides from FRAC Groups 3 or 9, such as Indar, Inspire Super, Procure/Trionic, Scala, or Vangard, during this period. Be sure to rotate FRAC Groups. Growers are highly encouraged not to use the FRAC Group 7 fungicides during this time period; these fungicides are best saved for peak apple scab pressure, which is from pink through petal fall.

From Pink through Petal Fall

Scab spores will start to peak (the maximum number of available spores dispersing from the overwintering leaves) beginning late pink and will remain high through approximately late petal fall. In our experience with monitoring scab spore dispersal from overwintering leaves, available scab spores remain high (more than 10,000) for approximately two weeks (from pink through petal fall).

During this time, it is best to use FRAC Group 7 (SDHI) fungicides, such as Aprovia, Fontelis, Luna Sensation, Luna Tranquility, Merivon, Pristine, or Sercadis, and tank-mix with a broadspectrum fungicide. Limit FRAC Group 7 fungicides to two applications during this period of high disease pressure. A maximum of four complete applications are allowed per year for FRAC Group 7 fungicides. Save two FRAC Group 7 fungicide sprays (if possible) for the end of the season when Luna Sensation, Merivon, or Pristine should be applied in order to mitigate late season and storage fruit rots.

From Post Petal Fall through Second Cover

Although the number of overwintering scab spores drastically decreases after petal fall, spores are still available and can wreak havoc, especially if conditions favorable for disease are present. During this time, use products from FRAC Group 3 and 9, such as Inspire Super, Indar, Rally, Procure/Trionic, Scala, or Vangard, plus a broad-spectrum fungicide. One recommendation is to use an EBDC through first or second cover and then switch to captan for the later summer cover sprays. Use products that may have a long PHI (such as Scala) earlier rather than later. These products could also be used in rotation with the FRAC Group 7 fungicides that are used from pink through petal fall.

Mitigating Fungicide Resistance for Brown Rot

Many factors influence brown rot development. During dormancy, removal of brown rot blossom blight cankers and fruit mummies will decrease the number of available spores during the season. Green fruit are not susceptible to infection by the brown rot pathogen. However, immature fruit that are not properly pollinated or become injured can become infected and begin to rot. Remove any infected green fruit and drop them to the ground. Near harvest, as fruit are maturing, drop any rotting fruit to the ground to prevent fruit from becoming mummies, thereby reducing overwintering inoculum for next year.

Bloom through Cover Sprays

The relative efficacies of current fungicides available are listed. Depending on disease conditions during bloom, one or two sprays will be needed for protection from blossom blight caused by the brown rot pathogen. Research from Rutgers has shown that captan cover sprays will adequately drop the number of available spores that could cause disease when harvest nears.

Preharvest Sprays for Brown Rot

If frequent rains continue throughout the summer and harvest season, then a three-spray preharvest program is highly recommended. The recommended timing for this program is 18 days, nine days, and one day preharvest, with a final captan cover spray at 28 days preharvest. Note that the final preharvest spray can be applied immediately before the first picking, or alternatively between the first and second picking; the idea is to provide protection throughout the handling process. Of course, the fungicide used at this time must have a zero- or one-day PHI and appropriate REI. For resistance management reasons, a minimum of two different chemistries should be applied to each cultivar block (alternated). However, use of three different chemistries is strongly recommended given that some of these chemistries are rated as high risk for development of resistance. An excellent three-spray program that utilizes all three chemistries is Gem, Indar, and Fontelis. For those fungicides composed of two active ingredients, simply alternate with the third chemistry. As you progress through the harvest season spraying different cultivar blocks, simply continue with the rotation.

Cedar Apple Rust

Only a brief part of the life cycle of the cedar apple rust fungus is spent on apple trees. Infection of apple leaves or fruit occurs between the pink and first cover spray periods. The cedar apple rust fungus survives 19 months or longer on red cedar. The contact between the fungus and the fungicide applied to apples is relatively short, reducing the potential for resistance to develop. If a resistant cedar apple rust fungus does develop, it must also survive on red cedar. Therefore, resistance of the cedar apple rust fungus to any fungicide is not likely.

Summer Diseases on Apple

Although resistance has not been reported for fruit rots or sooty blotch and flyspeck, it is important to be proactive by rotating fungicides and tank mixing with a broad spectrum chemical when controlling these diseases.

Biological Pest Control in Citrus Fruits

Citrus crops, mainly sweet oranges, provide the main part of the world's essential oil production. Important quantities of essential oils are also extracted from other citrus cultivars like tangerines, lemons, grapefruits, limes, and grapefruits. At the same time, the citrus ecosystem, offers ideal conditions for the proliferation of numerous pests and diseases. The use of chemical pesticides is the most common method of controlling the citrus pests. However, there are numerous problems associated with the misuse of pesticides, so increasingly the industry, including citrus producers, processors and end users, are driving the demand for chemical-free products. To meet this demand, non-chemical pest control strategies are required, needing to be both efficient and environmentally friendly as well as guarantee similar production yields. Biological pest control, alone or as a part of an Integrated Pest Management (IPM) strategy, represents such an alternative to chemical pesticides.

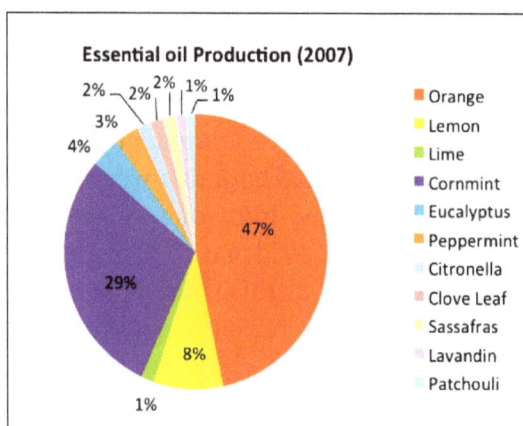

Figure: Origin of the top essential oils

Problems Associated with Pesticide use

Citrus pest control most commonly involves the use of synthetic insecticides. However, the abuse and misuse of synthetic insecticides has resulted in serious environmental problems and human health risks. Pesticide residues have been found in various parts of the environment, including humans, and are considered to be one of the main causes of the observed loss of biodiversity in agro-ecosystems, an undesirable side effect. At the same time, adverse health effects arise for consumers from food contaminated with pesticide residues and farm workers through pesticide exposure. Finally, the continuous use of pesticides has resulted in resistant pest populations, leading to pest control failures with severe economic consequences.

Pesticide residues are found principally in the peel of citrus fruits. Consequently, they are concentrated in the essential oils that are extracted from the peel.

In a recent study by Di Bella et al. 2010, the presence of organophosphorus and organochlorine pesticides in essential oils extracted from citrus originating from various countries (120 samples from Italy, and 70 samples from Brazil, Argentine, South Africa and Spain) were investigated. According to the results, pesticides from both categories were found in the majority of the samples examined. The largest quantities of pesticides were found in essential oils from Brazil and Spain. The presence of these pesticides in the citrus essential oils is probably associated with the insecticides used for pest control.

Given that citrus essential oils are used in the food, pharmaceutical and cosmetic industries, contamination with pesticides is undesirable. Therefore, alternative pest control methods able to protect the crop efficiently and at the same time reduce the presence of pesticides in citrus essential oils are of great importance.

Biological Control: An Alternative Method for Pest Control

Biological pest control is an environmentally safe method that represents a sound alternative to conventional chemical control.

Biological control involves the use of one or more types of beneficial organisms, to reduce the numbers of another type of organism or alternatively the use by humans of parasitoid, predator, pathogen, antagonist, or competitor populations to suppress a pest population, thereby making the pest less abundant and damaging than it would be in the absence of these organisms.

The provision of biological control is a mature and continuously growing industry, with more than 30 years of functioning at the highest professional level with well-developed mass production systems, quality control programs, a research sector, specific shipment and release methods and specialized advice for farmers. The plant protection market has an annual value of $35 billion out of which biologically based control methods (semio-chemicals, microbials, beneficial insects, etc.) account for $600-700 million. More than 170 species of invertebrate natural enemies are produced and sold worldwide for use in augmentative biological control of more than 100 pest species.

There are different types of biological control: classical biological control is used mostly against pests of foreign origin and involves the introduction of specialist natural enemies from their homeland. The objective is long-term, i.e. to establish populations of these natural enemies to attack the pest and to reduce its numbers. Special care should be taken in order to avoid possible non-target effects.

Augmentative biological control increases the populations of the natural enemies by means of massive releases of agents purchased from bio-factories. Augmentative biological control may be either inoculative or inundative. In the inoculative approach, the natural enemies are introduced in small numbers early in the season. These natural enemies establish, reproduce and feed on the pest and therefore they can suppress the pest populations. This approach is commonly used in greenhouse crops against various pests such as whitefly and aphids. The inundative approach involves mass releases of commercially reared natural enemies. This approach is used for rapid pest suppression, when the population of natural enemies is expected to be low and insufficient to control the pest.

Finally, conservation biological control involves the manipulation of the agricultural environment to preserve and enhance the existing populations of natural enemies. Manipulation may include the provision of alternative food sources for natural enemies such as nectar, pollen, alternative prey, or shelter for overwintering sites.

Integrated Pest Management (IPM)

Moreover, biological control is the cornerstone for IPM, an ecosystem-based strategy that focuses on long-term prevention of pests or their damage through a combination of techniques such as:

- Biological control.

- Cultural methods (habitat manipulation, modification of cultural practices, and use of resistant varieties).

- Pesticides (only after monitoring indicates they are needed and applied in a manner that minimizes risks to human health, beneficial and non-target organisms, and the environment).

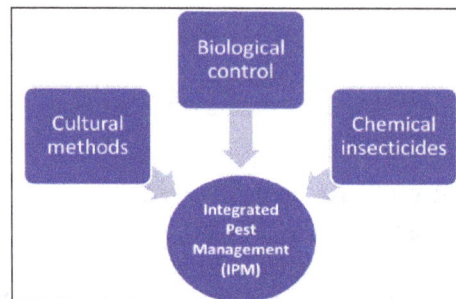

Pest control methods combined in an Integrated Pest Management strategy

Drivers for Change from Chemical to Biological Pest Control

There is an increasing tendency for change from chemical to biological control driven by the following factors:

Economics: It is generally assumed by growers that pesticides are cheap. This is because the benefits from pesticide use are based on the direct crop returns. Indirect costs such as environmental pollution and human health problems are not included in the pesticides' price and are usually paid by the society. If these costs are included in the comparison between chemical and biological control, the latter turns out to be the more economical solution. Moreover, the benefit/cost ratio and the developmental costs are much more favorable for biological control than for chemical control.

The change from chemical to biological control is additionally market driven; consumers, and consequently supermarkets, demand products free of pesticide residues. Finally, the preference for biological control is increasing as a result of the loss of chemical options available for pest control. This is firstly due to the resistance developed by many pests and secondly due to recent legislation. In particular the European Union Directive 2009/128/EC, aims to reduce the risks and impacts on human health and the environment related to the use of pesticides by promoting the use of alternative pest management methods.

Biological Pest Control in Citrus: Benefits for Essential Oil Production

A study conducted by Verzera et al. 2004, demonstrated that various advantages exist for essential oils that are extracted from fruits grown without the use of pesticides. The aforementioned study compared the presence of organophosphorus and organochlorine pesticides in sweet orange and lemon oils obtained from fruits grown under biological techniques with those obtained from fruits grown traditionally i.e. with insecticides used for pest control. Regarding sweet orange, various insecticides were detected in the traditional oils whereas the biological oils were almost free of insecticides: 90% less. Likewise, lemon oils extracted from biological fruits contained approximately 80% fewer insecticides than the oils extracted from those grown traditionally.

Most importantly, in the same study, the composition of the traditional and biological oils was also examined. The most relevant differences found were related to the content of aliphatic and terpene aldehydes. Biological fruits contained higher amounts of aliphatic aldehydes (average content 0.79%) than oils obtained from fruits coming from traditional orchards (0.22%). Moreover, traditionally grown fruits contained on average 2.41% of "terpene aldehydes" while biological oils containe 2.90%. As the authors explain, with the term 'terpene aldehydes' they refer mainly to neral and geranial, the citral substances that determine the distinctive lemon oil odor. Interestingly, the citral content largely determines the market value of the oil. The content of aliphatic and terpene aldehydes is an index of quality for orange and lemon oils respectively, and moreover, largely determines the market value of the oil. Therefore, the above results suggest that pesticide-free oils are of higher quality and greater market value.

Biological Control of Tephritidae

The biological control efforts against fruit flies of the genus Tephritidae have been extensive over the past half century, a thorough review being given in Clausen. However, as it becomes increasingly apparent that the Mediterranean fruit fly, Ceratitis capitata (Wiedemann), and Mexican fruit fly, Anastrepha ludens (Loew) pose a continued threat to California's agriculture through periodic invasions of our borders, there is an urgent need to consider the application of alternative methods to chemicals in eradication and control programs. The implementation of effective biological controls at the sources of invasion as well as within the state boundaries where breeding may occur, offers an environmentally sound, non-polluting alternative. There is an urgent need to (1) search for, procure and initially evaluate natural enemies of Mediterranean and Mexican fruit flies fruit fly in natural ranges in central Africa and southern Mexico (parasites, predators and pathogens); (2) introduce and study foreign natural enemies in the adult stage, and evaluate their respective effectiveness under field conditions in

Hawaii, southern Mexico, and if applicable, California; (3) attempt development of a mass production scheme of resident California fruit flies (e.g. walnut husk fly) to serve as acceptable hosts for Mediterranean fruitfly natural enemies for use in laboratory study and periodic colonization efforts in infested areas of California, and (4) to test the feasibility of building a culture bank of Medfly and Mexican fruit fly natural enemies on resident California fruit flies for use in conjunction with other eradication and control methods (e.g. sterile-male releases, adult fly baiting) during periodic invasions of these pests and in anticipation of their possible permanent establishment in the State of California.

The fruit flies of the family Tephritidae constitute a group of agricultural pests of worldwide importance, as they attack a wide range of fruits and vegetables. The most important are the several species of Dacus and Ceratitis, which occur in many countries of warm temperate and subtropical climates; Anastrepha, an American genus occurring from Mexico and the West Indies through South America; and Rhagoletis, with a more restricted host range, occurring in the north temperate region. The Mediterranean fruit fly, although eradicated periodically from the state of Florida where it had "peninsular" distribution, and recently from California where it repeatedly reappears, is presently firmly established in southern Mexico. There it has been temporarily contained by a massive sterile-male and parasite release effort by the U. S. Department of Agriculture. The appearance of Anastrepha in southern Baja California during the past two years suggests that it may eventually move north and pose a continuous threat along the Mexican border. Another chronic threat has been the permanently established population in the Hawaiian Islands, from which periodic accidental invasions of California are thought to occur. Recently, Carey & Dowell, Greathead & Waage, Gilstrap et al, Wharton and Wong and Ramadan have noted that further biological control efforts are definitely justified against fruit flies.

The need for investigation into the biological control of fruit flies in Hawaii, Mexico and California is ever more important as it becomes recognized that insecticides, although offering expedient and predictable results under certain conditions, are often inadequate and at least perceived as dangerous, if not physically dangerous to wildlife and humans alike. As problems involving insecticidal residues and insect resistance to chemicals continue to increase, many programs directed at the control of fruit flies must ultimately be modified with increased dosages and costs to such an extent that they invariably arouse the concern and ire of naturalist and conservationist organizations. A case in point is the fire ant eradication program. By 1959 extensive damage to wildlife and domestic animals had positively been attributed to the effects of several insecticides used in the program. Fire ant control was finally declared unsuccessful in 1960, and in some states, fire ant numbers were actually reported to have increased since the eradication program began. Presently, a new effort to control fire ant is being attempted with natural enemies imported from Brazil and Argentina.

Some investigators believe that the Mediterranean and Mexican fruit flies are already permanently established in California and that unless the current eradication effort is greatly increased, it is just a matter of time before at least one species, Medfly, will spread throughout the state. The malathion baits currently in use against them may not be potent enough for fast eradication, as it is recognized that Medflies will not eat the bait unless that is the only substance placed in their cages. Under outdoor conditions they may prefer to seek out clean ripening fruit.

Specific Examples

Mediterranean Fruit Fly: The Mediterranean fruit fly is a major pest throughout the Mediterranean region, portions of Africa, the Middle East, Central and South America, Mexico, and Hawaii, and has become established in Australia. In France, it is able to persist from year to year only in areas bordering the Mediterranean, yet survival is reported in Austria, where severe winters, with continuous frosts, can cause up to 90% mortality of the pupae. Although parasitic insects have been imported against it, 95% of the species were obtained from areas outside the fly's accepted native range in central Africa and Madagascar. However, some reductions in infestations are attributable to natural enemy activity in the invaded areas, especially when parasitoids are mass released as biotic insecticides.

The medfly was first described fin 1824 and was first noted as a pest in citrus in 1829 from shipments of oranges to England from the Azores. The fly spread throughout the world over the next 100 years and was continually noted as a destructive pest wherever it was found. The first program for the biological control of medfly was undertaken by the government of Western Australia in 1902 with the engagement of George Compere to search for natural enemies and to determine the aboriginal home of the medfly. Unfortunately Compere was never able to acertain the aboriginal home nor did he establish the parasitoids he collected from India, Sri Lanka and Brazil in Western Australia.

The medfly invaded Hawaii in 1910 and soon thereafter the Board of Commissioners hired Filipi Silvestri to again search for natural enemies of this fly. It was determined by experts of the day that collections should concentrate in Western Africa. Therefore, Silvestri traveled for eight months through West and East Africa and South Africa. He found only six specimens of the medfly on the entire journey, but reared many parasitoids from other fruit-infesting tephritids collected along the way. He managed to establish four species in Hawaii: Opius concolor Szepligeti, Biosteres tryoni (Cameron), Coptera silvestrii Kieffer and Dirhinus anthracina Walker. Two more missions over the next 30 years were sent out in hopes of obtaining parasitoids, but only Tetrastichus giffardianus Silvestri and Biosteres fullaway (Silvestri) were established.

Other biological control programs were undertaken in several countries where the medfly was firmly established, but these programs have not been well documented, and the extent of control of any of the parasitoid species is virtually unknown, the notable exception being Hawaii. Even in Hawaii control was never noteworthy and the medfly proglem was finally overshadowed by the introduction of Dacus dorsalis Hendel. For North America the answer to the medfly invasions starting in 1929 was complete eradication by means of fruit stripping and poisoned bait sprays.

The success of these early and subsequent biological control programs against medfly has been variable. In Hawaii, a cooperative biological control program initiated in 1948 involved the release of 32 entomophagous species to compat both medfly and the oriental fruit fly. Three parasitoids, Biosteres longicaudatus (Ashmead), B. vandenboschi (Fullaway), and B. oophilus (Fullaway) became widespread and abundant. During 1966-1968, parasitization of the medfly and the oriental fruit fly was high (ca. 70%); it was mainly due to the egg-pupal parasitoid, B. oophilus. During 1978-1981, Biosteres oophilus was still the predominant parasitoid as it accounted for ca. 80% of the total parasitization. Occasionally the larval-pupal parasitoids, Biosteres longicaudatus and B. tryoni (Cameron) achieved a parasitization of 32 and 8%, respectively. Extensive fruit collections done between 1949-1985 showed that Jerusalem cherry, coffee and peach were among the most important hosts of the medfly. These fruits yielded more than 100 larvae/Kg of infested fruits. The fruits that yielded a high number of medfly larvae were elliptical to spherical and yellowish to reddish. They had a diameter of 1-7 cm and a weight of 1-30 grams. Most of these hosts belonged to five plant families: Myrtaceae, Rutaceae, Rosaceae, Sapotaceae and Solanaceae.

In Costa Rica a classical biological control program was initiated in 1955. During 1979-1980 parasitoids were collected from <10% of C. capitata populations. These were two introduced braconids, B. longicaudatus and B. oophilus, and two indigenous cynipids, Ganaspis carvalhoi (Dettmer) and Odontosema anastrephae (Borgmeier). An exploration for medfly parasitoids conducted in West-Central Africa during 1980-1982 showed that C. capitata occurred in low frequency in coffee plantations in Cameroon. Parasitization of tephritids in coffee by braconids ranged from 10-56%. In Guatemala infestation of C. capitata was serious in coffee and tangerine. The rest of the fruits were mainly infested by Anastrepha spp. Parasitization rate of C. capitata and Anastrepha spp. was low, ranging from 0.04 to 7.95%. The most common parasitoids recovered from both flies were B. longicaudatus and Doryctobracon crawfordi (Viereck).

The behavior of the ectoparasitoid Muscidifurax raptor (Girault & Sanders) in searching for the potential host C. capitata pupae was analyzed under laboratory conditions. The searching efficiency of M. raptor females decreased with increasing density. The proportion of avoidance of superparasitism was 0.615. The response to a high parasitoid density was to increase the proportion of males in the progeny, as males searching for mates interfered and decreased the searching activity of the females. The medfly was susceptible to the Mexican strain of the nematode Steinernema feltiae. Emerging adults and pupae were not susceptible to the nematode, but the third instars (prior to pupating in the soil) suffered high mortalities (50-90%) when exposed to high nematode concentrations (150,000 - 500,000 nematodes/cup). Field exposure of mature larvae to a dose of 500 nematodes/cm^2 yielded high mortality of C. capitata. In addition to the hymenopterous parasitoids and insect pathogenic nematodes, arthropod predators such as ants could play an important role in reducing fruit fly populations. Under laboratory conditions, the argentine ant, Iridomyrmex humilis (Mayr) caused 50% mortality of medfly pupae after a 10 min. attack. However, ant predation could be important only in localized areas; it is not adequate to regulate medfly populations.

Typically, the most effective natural enemies of an insect occur in regions where the pest evolved. The natural range of Mediterranean fruit fly is the sub-Saharan central African region, including the Island of Madagascar. Although no information is available from Madagascar, a number of promising natural enemies have been discovered in Central Africa. However, because of technological difficulties associated with transportation and culture, only two species attacking Ceratitis capitata have been successfully translocated out of central Africa. A concentrated effort to locate natural enemies there might yield the kind of species capable of regulating this pest at low densities, as it has been known to be rare in that general region since the early 1900's. We believe that parasitic Hymenoptera are the most effective natural enemies of Mediterranean fruit flies. At least six species of fruit flies in the genus Ceratitis are known from central Africa, and numerous parasitic Hymenoptera have been reported active on them at very low host densities. Entomologists in California have not tested these because the Mediterranean fruit fly has been quarantined. Therefore, promising species of natural enemies for Medfly might be found among these related species. Also, there has been no concentrated effort to locate disease organisms, such as viruses, bacteria and fungi, which might prove invaluable in eradication campaigns.

Mexican fruit fly: Some of the natural enemies of oriental and Mediterranean fruit flies have shown activity on Anastrepha spp. in southern Mexico, and may be influential in partial biological control of that species. However, there have been no formal attempts to obtain natural enemies from other areas where different species of Anastrepha occur, such as South America.

Two species of parasitic insects are already proven and available as biotic insecticides (augmentive releases) against Medfly. These are Diachasmimorpha longicaudata and D. tryoni, which have been used with some success in Mexico and Hawaii. The use of these parasites in lieu of malathion during the establishment phase of specific natural enemies from central Africa, would greatly aid their survival and while providing some economic control of Medfly.

Table Known parasitic species attacking fruitflies of the genus Ceratitis in their natural range in Central Africa.

Parasite species	Host stage attacked
Biosteres caudatus Szepligeti	larva
Diachasma fullawayi Silvestri	larva
Diachasmimorpha longicaudata (Ashmead)	larva
Diachasmimorpha tryoni (Cameron)	larva
Dirhinus ehrhorni Silvestri	pupa
Dirhinus giffardii Silvestri	pupa
Ganaspis sp.	larva?
Halticoptera sp.	larva?
Hedylus giffardii Silvestri	larva?
Hedylus sp.	larva?
Galesus silvestrii Kieffer	pupa
Microbracon celer (Szepligeti)	larva?
Opius humilis Silvestri	larva
Opius inconsuetus Silvestri	larva
Opius perproximus Silvestri	larva
Opius n. sp.	larva
Spalangia afra Silvestri	pupa
Tetrastichus dacicida Silvestri	larva
Tetrastichus giffardii Silvestri	larva
Tetrastichus oxyurus Silvestri	larva
Tetrastichus n. sp.	larva
	larva

Misc Fruitflies: Several species of Rhagoletis are very important pests of cultivated cherries in North America and Europe, with some species having been considered as subjects for biological control, despite the low economic threshold. Infestation rates of less than 0.2% are currently required for commercial marketing of cherries in the United States. Four species of parasitoids associated with the Oriental fruit fly, Dacus dorsalis Hendel, were introduced against such fruit flies. These included Opius longicaudatus compensans (Silv.), Opius longicaudatus farmosanus (Full.), Opius oophilus Full, and Opius longicaudatus novacaledonicus Full. These parasitoids were introduced from Hawaii and released against Rhagoletis indifferens Cueran and Rhagoletis fausta Osten Sak in Oregon and Washington in the 1950's. However, none became established probably because they all originated in tropical regions. A parasitoid of R. cerasi, the European cherry fruit fly, was imported against the eastern cherry fruit fly, R. cingulata Loew during 1959-64 in New Jersey, without successful establishment. Other species including Biosteres sublaevis Wharton, Coptera occidentalis and Phygadeuon wiesmanni are under investigation in California and Oregon.

References

- Pests-and-diseases-of-fruit-trees: countryfarm-lifestyles.com, Retrieved 21 April, 2019
- Peach-leaf-curl-symptoms-and-control, plant-diseases, plants: biologydiscussion.com, Retrieved 2 February, 2019
- BrownRotStoneFruits: apsnet.org, Retrieved 30 March, 2019
- Orchard-ipm-natural-enemies-biological-control-in-orchards: psu.edu, Retrieved 28 January, 2019
- Tree-fruit-disease-toolbox-fungicide-resistance-management: psu.edu, Retrieved 8 March, 2019
- Biological-pest-control-in-citrus, alternative-to-chemical-pesticides, benefits-for-essential-oil-quality: researchgate.net, Retrieved 18 May, 2019

Biocontrol of Diseases of Seed

Diseases of seed can affect seeds or seedlings and are generally responsible for thin stands and poor emergence. Such diseases cause damping-off, fading out or seedling blight. The topics elaborated in this chapter will help in gaining a better perspective about the different types of diseases of seed as well as their treatments.

Seedling Diseases

Seedling diseases create problems for those tying to establish grass from seed in the fall. Pythium, Rhizoctonia, Helmintho-sporium, Curvularia and Fusarium all contribute to a disease complex causing damping-off, fading out or seedling blight. The disease attacks seemingly healthy vigorous stands of seedlings and kills the young plants in patches. Seedling diseases are especially damaging during adverse weather conditions-unusually warm periods in the fall, continued wet conditions, or cool, wet periods in early fall. Planting too early (or too late) also increases the incidence of seedling diseases.

The terms "damping-off" or "seedling blight" are used to describe several seedling diseases. The diseases may be incited by fungi or by environmental conditions. Excessive or inadequate soil moisture, cool soil temperatures, humid and unseasonably warm temperatures, saline soils, compacted soils, or other environmental stresses can lead to damping-off or seedling blight.

Symptoms

Damping-off and seedling blight can occur before seedling emergence (preemergence) or after seedling emergence (post-emergence). Pythium incited damping-off is characterized by a high order of preemergence killing of seedlings. In this case, the deterioration process begins soon after the seed coat is broken.

In the case of postemergence damping-off, seedlings emerge above the soil and begin deteriorating at the soil level. As the deterioration progresses upward the seedlings appear watersoaked. As the tissue collapses, the seedlings shrivel and turn brown. Pythium, Fusarium, Rhizoctonia, Helminthospor-ium and Curvularia may all cause postemergence damping-off.

The effect of these seedling diseases is a significantly reduced stand of grass that usually requires replanting. If the disease outbreak is treated in time, the stand may only be thinned by the attack.

Seed Borne Pathogen

Seed borne pathogens causes diseases at various stages of crop growth from germination of seed up to crop maturity and heavy losses have been observed, caused by seed borne pathogen in various crops. Seed borne pathogens causes seed and seedling rots, i.e. pre- and post-emergence losses, diseases at various stages of crop growth like root rot, stem rot, fruit rot, wilt, blight, leaf spot etc. Influence the crop stand and ultimate yield. Therefore, the good seed must not be affected by any seed borne pathogen. Pathogen free seed is a factor which needs the maximum attention of farmer for an increase crop production. Thus, detection of plant pathogen from seed and their estimation and management is very important for agriculture production/yield.

Ways of Infection of Seed Borne Pathogen

Externally Seed Borne Pathogen

The seed inoculum in such cases is superficial and confined to the surface of seed, usually as adhering propagules, e.g. spores sclerotia, mycelium, bacteria, nematodes, virus particles etc. Contamination of seed surface, especially by fungi is often detectable by direct observation under microscope or by examining seed washing.

Internally Seed Borne Pathogen

The inoculum lies with the tissues i.e., this pathogen are carried inside the seed, usually as adhering by vegetative cell, spores, pycnidia, nematodes or virus particles. Dry seed may look perfectly healthy when examined under a binocular microscope and no signs of infection. Seed borne pathogen established with seed coat, testa, pericarp, endosperm and embryo.

Concomitant Contamination

The inoculum is present as contamination mixed with seed in the from of infected debris fungal sclerotic, bacterial ooze, nematode cysts, infected soil particles etc. Such contamination is difficult to detect.

Transmission of Seed Borne Pathogens or Disease

Seed plays a vital role in the transmission of pathogens directly or indirectly. It is essential to understand precisely how the organisms are associated with the seed and get transmitted. The type of pathogen transmitted includes seeds of plant (phanerogamic plant parasite), nematodes, fungi, bacteria and viruses.

Plant pathogens are seed transmissible by:

1. Adhering to seed surface,

2. Becoming internally established with in the seed and

3. Accompanying the seed lot as infected plant debris, soil clad or adhering to containers or otherwise.

Detection Techniques of Seed Borne Pathogens

Several methods have been developed to detect seed borne micro flora. The method of detection may be general or specific for individual pathogen.

The selection of seed testing method for a particular study is based on certain Objectives:

1. Testing for quarantine purposes.

2. Testing for national seed certification schemes.

3. Testing for evaluating the planting value of the seed.

4. Testing for storage fungi.

Generally, according to the International seed Testing Association (ISTA) until and unless otherwise stated, a minimum of 400 seed should be tested for each sample.

Prevention Methods against Seed Borne Pathogens

An outline of the measures of prevention of diseases due to seed borne pathogens, as modified from Baker, is enumerated below:

- Management Practices

  1. Seed source-pathogen free seed.

  2. Selection of seed-production area and season when and where the seed is not likely to carry pathogens.

  3. Seed-field inspection.

  4. Seed certification.

  5. Quarantine.

The measures 1st to 4th attempt elimination and 5th is for avoidance.

- Cultural Practices in Seed-Production Fields

  1. Sowing methods, e.g. deep sowing and planting.

  2. Pathogen control in seed field: Control of weed hosts, production including pre-harvest earhead spray with fungicides as in wheat or rice if there is rain.

  3. Avoidance of overhead watering: Ditch irrigation, rather than sprinkler irrigation is favourable for seed crops. This is particularly true in semiarid areas where foliage would otherwise remain uninfected.

  4. Harvesting method: Delaying harvesting, and various cares taken during harvesting.

  5. Eradication of infected host plants: Applicable when a disease is newly introduced in an area.

6. Ageing of seed to utilize the phenomenon that some seeds remain viable for a period longer than the period of survival of the pathogen, as in cucurbits.

7. Treatment of field soil-Occasionally effective.

The measures are against the directed reduction of established inoculum. The measures 3rd arid 4th also reduce inoculum build up.

- Curative or Eradicative Measures for Seed already Contaminated

    1. Seed indexing.

    2. Separating procedures.

    3. Chemical seed treatment: Seed treating chemicals with low mammalian toxicity such as antibiotics (aureofungin, blasticidin, etc. against rice blast and brown spot), organic sulphur, systemic fungicides and their combinations (thiram+carboxin; thiram+bavistin, etc.) have been developed to replace organic mercury.

    4. Thermotherapy of seed.

The above measures are intended for reduction of established inoculum.

Breeding for Resistance against Seed Transmission

That the amount of primary inoculum should be limited in seed should be the chief objective of control of the diseases due to seed borne pathogens. Thus, breeding disease resistant varieties is likely to be a very successful measure against these diseases. The integration of different measures would be necessary for the evaluation of a recommendation for any specific case.

Soybean Seed and Seedling Diseases

Seed and seedling diseases of soybean are common and significant problems. They can decrease plant populations that result in replanting and production losses. Several different pathogens can cause these diseases, and the most common tend to be Fusarium, Rhizoctonia, Phytophthora, and Pythium. They can kill and rot seeds before germination or cause seedling death. They are most common when soil is very wet in the first few weeks after planting and in heavy, poorly-drained soils.

Symptoms

Evidence of seed and seedling diseases is usually seen when seedlings don't emerge or they die or are stunted. Infection and damage prior to emergence is common but difficult to identify. Factors other than disease can also cause these problems, so it is important to look closely to determine the cause(s). Different seedling diseases may cause different symptoms, but similarities can make them difficult to distinguish.

Phytophthora can attack and rot seeds prior to emergence, and can cause pre- and post- emergence damping off. It produces tan-brown, soft, rotted tissue. At the primary leaf stage (V1), infected stems appear bruised and soft, secondary roots are rotted, the leaves turn yellow, and plants frequently wilt and die.

Pythium can attack and rot seeds and seedlings prior to emergence and can cause post-emergence damping-off under wet conditions. The characteristic symptom of most Pythium infections is soft, brownish-colored, rotting tissue. Thus, Pythium causes symptoms similar to Phytophthora in seedlings, and can only be distinguished by laboratory examination. Although Pythium causes most damage to seeds and seedlings, roots of established plants can be rotted and plants may be stunted.

Rhizoctonia can damage seeds and plants prior to or after emergence. In seedlings and older plants, a firm, rusty-brown decay or sunken lesion on the root or on the lower stem is a characteristic symptom. The infections can be superficial and cause no noticeable damage, or they can girdle the stem and stunt or kill plants.

Fusarium is also a common pathogen that can damage seeds and seedlings. It causes light to dark brown lesions on roots that may spread over much of the root system and may appear shrunken. Fusarium may attack the tap root and promote adventitious root growth near the soil surface and may degrade lateral roots.

Conditions and Timing that Favor Disease

General conditions that promote seed and seedling disease diseases include wet, poorly-drained, and compacted soils. However, the different pathogens have different optimal conditions. For example, Phytophthora and Rhizoctonia are favored by wet and warm soils, whereas Pythium is typically favored by wet and cool soils. Seed and seedling diseases may be enhanced by slow germination and growth of soybeans, poor quality seed, and plant stress.

Causal Pathogen

- Most of the important seed and seedling diseases of soybeans are soilborne, although the role of seedborne inoculum is not well understood. They are spread primarily with soil and water, but can also be spread with dirty seed. Most are probably indigenous soil inhabitants in soybean fields. They can generally survive for long periods in the soil, often associated with plant debris.

- The most important pathogens involved in seed and seedling diseases of soybeans in the Midwestern U.S. are fungi or fungal-like organisms: Fusarium spp, Rhizoctonia solani, Phytophthora sojae, and Pythium spp. In some situations and locations, other soilborne or seedborne pathogens that may be important include Macrophomina, Colletotrichum, and Phomopsis.

- Phytophthora sojae is a fungal-like pathogen that survives in soil for up to five to 10 years in association with decomposed soybean tissues. Soybean is the only known crop host for this pathogen. It is favored by saturated, warm soil.

- Pythium is a soilborne, fungal-like pathogen: Several different species damage soybeans. The various species of Pythium that infect soybean have a wide host range that can include corn and many other crops. Pythium tend to be favored by cool and soil, but some species may do more damage in warm soils.

- Rhizoctonia solani is a common pathogen with a wide host range: The most common strains of this pathogen (anastamosis groups, AG) that infect soybean are AG-2-2 and AG-4. AG groups can have different optimal conditions for infection.

- Fusarium seed and seedling blight of soybean is caused by a complex of different species that may prefer different conditions. For example, some species may prefer warm and dry soils and others may prefer cool and wet soils. Some Fusarium species may also have a broad host range that includes corn and wheat.

Cotton Seedling Diseases

Cotton seedling disease and seed rots are caused by several different species of fungi. Primary agents in North Carolina include Rhizoctonia solani (Soreshin), Pythium spp. (Root Rot), Fusarium spp., Thielaviopsis basicola (Black Root Rot), and Phoma exigua (Ascochyta gossypii). These pathogens are found everywhere cotton is grown, and colonize weak cotton plants.

A poor stand of cotton seedlings caused by post-emergence damping-off fungi.

Environmental Factors Influencing Seedling Diseases

Plants are more prone to infections by pathogens when stressed by an inhospitable environment, and environmental factors are very important in influencing the development of seedling diseases.

Seedling disease occurs more frequently under cool, wet conditions and more prevalent on sandy, low-organic-matter soils. Other factors, such as planting too deep, poor seed bed conditions, compacted soil, nematode or insect infestations, and misapplication of soil-applied herbicides, may increase the problem. Additionally, seedling diseases tend to be more severe in reduced tillage situations and when beds are absent. Planting on beds elevates the seed allowing for more rapid emergence and improved water drainage, especially after heavy rains. Plants are also more prone to attack by pathogens when insects or nematodes cause damage to the root and crown of the plant. Damage from thrips in particular can delay seedling development and enhance damping-off diseases caused by various fungi.

Symptoms

Rhizoctonia solani invades at soil level from time seedling emerges until growth around 6 inches after which stem becomes woody and infection is less likely to occur unless injured. Damage to the hypocotyl, called "soreshin", is derived from sunken reddish-brown lesion which girdles the hypocotyl causing the seedlings to collapse. Surviving seedlings may show stunted growth.

Pythium zoospores actively move in water. Commonly found in soils that are poorly drained or have been heavily saturated with water for several days. This pathogen causes seed rot and pre- and post-emergence damping off. Water soaked lesions appear and cause seedlings to collapse.

Fusarium species survive indefinitely in the soil as fungal overwintering structures, and colonize weak, damaged tissues. Infections by Fusarium species results in stunted growth and failure to thrive. Some strains will only display symptoms, usually nematode galling on lateral roots, when also infected with root knot nematode.

Thielaviopsis basicola infects seedlings at the epidermis and root cortex. Infected seedlings display a thinner dark brown to black discolored taproot. The pathogen leads to a loss of mycorrhizal symbionts and rotting of the cortex tissue. Thielaviopsis basicola rarely kills seedlings but does lead to stunted growth. As the plant matures new lateral roots and cortical cells develop and the taproot expands, shedding the dead, blackened cortical cells.

Phoma exigua (syn. Ascochyta gossypii) develops in periods of cold, wet weather and infects the main stem at a node. Cotyledons prematurely turn brown and die. Below wilted leaves Phoma can create a black canker. Streaks found in the stem don't usually extend to the roots like they do with infections by Fusarium species.

Sore shin on cotton seedlings.

Cotton seedling death caused by Phoma exigua.

Disease Cycle

Although seedling diseases can be caused by numerous pathogens, these pathogens are most severe when seedlings are slow to emerge. Cool, wet climates increase the susceptibility of cotton to these pathogens.

Rhizoctonia Solani

Rhizoctonia solani hyphae emerge and produce mycelium. The mycelia land on plant surfaces and secrete necessary enzymes to initiate invasion into host. Necrosis and sclerotia form in and around infected tissue. The pathogen is dispersed as sclerotia and can travel by wind, water, or soil movement. Rhizoctonia solani overwinters in plant debris, soil or host as sclerotia.

Pythium spp.

Pythium infects host tissue through an encysted zoospore germ tube and overwinter as an oospore. Disease becomes more prevalent in cold, wet weather.

Fusarium spp.

Fusarium spores begin to germinate in the presence of developing roots and root exudates. Pathogen enters through direct penetration or wounds and spreads through seedling by microconidia in the xylem. Microconidia clog the xylem tissue and prevent water and nutrients from traveling to the rest of the plant. Dead plants return spores to the soil and move through irrigation water. Chlamydospores can survive in soil and plant residues for years.

Thielaviopsis Basicola

Thielaviopsis basicola overwinter as chlamydospores which can survive in soil for many years. Spores begin to germinate when root exudates are detected and soil conditions are favorable. Thielaviopsis basicola development is most severe during cool, wet weather.

Phoma Exigua

Phoma exigua is most severe when night temperatures fall between 50 - 60 °F. The pathogen begins to attack seedlings from the time they emerge until they are about six inches tall. After the plant stems becomes woody further infection rarely occurs.

Damping off

Damping off affects many vegetables and flowers. It is caused by a fungus or mold that thrive in cool, wet conditions. It is most common in young seedlings.It can cause root rot or crown rot in more mature plants.

Seedlings infected by damping off rarely survive to produce a vigorous plant. Quite often a large section or an entire tray of seedlings is killed.

Visible damping off fungus growing on an emerging seedling.

Once plants have mature leaves and a well developed root system, they are better able to naturally resist the fungus or mold that causes damping off. There is a critical period of growth between planting and maturity when special care needs to be taken to protect sensitive seedlings.

A wide variety of vegetables and flowers can be affected by damping off. Young leaves, roots and stems of newly emerged seedlings are highly susceptible to infection. Under certain environmental conditions, damping off pathogens can cause root rot or crown rot in mature plants.

The fungi, Rhizoctonia spp. and Fusarium spp., along with the water mold Pythiumspp. are the most common pathogens responsible for damping off.

Identifying Damping off Symptoms

Seedlings fail to emerge from the soil. Cotyledons (the first leaves produced by a seedling) and seedling stems are water soaked, soft, mushy and may be discolored gray to brown. Seedling stems become water soaked and thin, almost thread like, where infected. Young leaves wilt and turn green-gray to brown. Roots are absent, stunted or have grayish-brown sunken spots. Fluffy white cobweb-like growth on infected plant parts under high humidity.

Causes of Damping off

All of the pathogens (fungi and molds) responsible for damping off survive well in soil and plant debris.

The pathogens can be introduced into the seedling tray in several ways.

- Pots, tools, and potting media that have been used in previous seasons and are not properly cleaned can harbor the pathogens.
- Spores of Fusarium spp. can be blown in and carried by insects like fungus gnats, or move in splashing irrigation water.
- Pythium spp. is often introduced on dirty hands, contaminated tools or by hose ends that have been in contact with dirt and debris.

Once introduced to a seedling tray, the damping off pathogens easily move from plant to plant by growing through the potting media or in shared irrigation water.

Garden soil often contains small amounts of the damping off pathogens. If you use garden soil to fill seedling trays, you could introduce the damping off pathogens that cause the disease into the warm wet conditions best for seed growth.

Mushy tan spots on these seedlings are signs of infection by damping off fungi that can be caused by over watering.

Seeds planted directly into the garden can also suffer from damping off. Disease is particularly severe when seeds are planted in soils that are too cool for optimal germination or when weather turns cool and wet after planting resulting in slow germination and growth.

The damping off pathogens thrive in cool wet conditions. And any condition that slows plant growth will increase damping off. Low light, overwatering, high salts from over fertilizing and cool soil temperatures are all associated with increased damping off.

Seedling Blight

The disease appears circular, dull green patch on both the surface of the cotyledon leaves. It later spreads and causes rotting. The infection moves to stem and causes withering and death of seedling. In mature plants, the infection initially appears on the young leaves and spreads to petiole and stem causing black discoloration and severe defoliation.

| Dead seedling | Spot on older leaf Leaf | blight symptom |

Pathogen

The pathogen produces non-septate and hyaline mycelium. Sporangiophores emerge through the stomata on the lower surface singly or in groups. They are unbranched and bear single celled, hyaline, round or oval sporangia at the tip singly. The sporangia germinate to produce abundant zoospores. The fungus also produces oospores and chlamydospores in adverse seasons.

Favourable Conditions

• Continuous rainy weather.

- Low temperature (20-25 °C).

- Low lying and ill drained soils.

Disease Cycle

The pathogen remains in the soil as chlamydospores and oospores which act as primary source of infection. The fungus also survives on other hosts like potato, tomato, brinjal, sesamum etc. The secondary spread takes place through wind borne sporangia.

Diseases of Corn Seed and Seedlings

Most corn seed and seedling diseases have several common generalized effects. Plants may fail to emerge due to seed rot or preemergence damping-off; growth may be slow, resulting in stunted plants; plants may appear yellow; wilting may occur; and plants may collapse due to postemergence damping-off. In some cases, infection may cause damage that persists into the growing season.

It can be difficult to distinguish environmental stress from some disease damage, but symptoms can help deduce whether pathogens are causing problems. Symptoms of seed and root infections include rotted seed; rotted seedlings (preemergence damping-off); leaf tip necrosis; stunting (sometimes with a mixture of short and tall plants); yellowing-reddening of older leaves; roots that are rotted, pruned-off, and discolored with firm or soft, brown-reddish to gray lesions or decay; poorly developed root systems; soft and discolored coeleoptile; leaf tip necrosis in streaks or patches; wilting seedlings; and discolored sunken and soft lesions on mesocotyl. Symptoms of foliar seedling infections include round to elliptical tan spots on leaves (holcus spot); necrotic, wavy necrotic streaks on leaves (Stewart's wilt); and oval lesions on seedling leaves (anthracnose).

A number of different pathogens can cause one or more of the symptoms of seed and seedling disease. Some of the more common genera of fungi or fungal-like organisms are Pythium, Fusarium, Rhizoctonia, Diplodia (Stenocarpella), Colletotrichum, and Penicillium. Their importance will depend in part on crop rotation, location, and soil conditions. Pythium is a widespread soil fungal-like pathogen that causes seed rot and seedling infections, and is favored by wet and cool soil conditions. Erwinia (Pantoea), which causes Stewart's wilt and Pseudomonas, which causes holcus spot, are two bacterial pathogens that can infect corn seedlings. Nematodes also may affect corn seedlings in some areas.

Various conditions, factors, and cultural practices favor infection by these pathogens. The following list is not complete, but covers common problems. Favorable factors include poor seed quality (such as cracked seed and infected seed), soil temperatures at planting below 50 to 55 deg F, wet soil conditions, soil compaction, slow emergence and growth, fertilizer burn, improper use of pesticides, injury from herbicides, soil crusting, high temperatures, high populations of flea beetles (Stewart's wilt), and sandy soils (nematodes). The obvious theme here is a combination of conditions that favor the pathogens and stress the corn seed and seedlings.

Fortunately, in most normal springs, seed and seedling diseases cause minimal damage in most fields. Other than using fungicidal seed treatments, no special management tactics need be put into effect. Most management tactics are obvious from the list above: follow good agronomic practices and try to avoid those conditions and factors that favor seed and seedling diseases. In addition, crop rotation and practices that minimize corn residue may be of value. Most seed corn is treated with fungicides. The treatments used have distinct efficacies against various pathogens. The primary fungicidal seed treatments used are of two main groups. The first group targets Pythium; some examples are ApronXL, Allegiance, and Apron. The second group targets the true fungi; two examples are Captan and Maxim. Other products may also be available that offer similar efficacy, and products are available as mixtures of the active ingredients from both of these fungicidal groups.

Fungal Diseases of Seedling

Seedling disease can be caused by a variety of different fungi harming plants in the early growth stage. Seedling diseases are often a complex of two or more different fungi infecting a plant. But seedling disease is also a complex because it may appear at different stages of the young plant's growth.

Whatever fungi or what stage of the seedling is affected, nearly all crops propagated from seed face serious loss due to seedling diseases.

The effects of the seedling disease may appear as a seed rot (pre-emergence damping off), seedling decay before the seedling emerges (pre-emergence damping off), seedling decay after the seedling emerges (post-emergence damping off), or seedling root rot (root pruning). In pre-emergence damping off, the seed rots and never germinates, or the seed germinates but the seedling succumbs to the disease and dies before it can emerge. Post-emergence damping off is when the seedling emerges and then dies soon after. Whether the seedling disease is a pre- or post-emergence damping off, the results are thin or uneven plant stands. The overall effects of seedling root rots are often subtle, causing reduced seedling vigor.

The fungi that cause seedling diseases are common pests to most all seeded herbaceous plants. The most common are Pythium species and Rhizoctonia solani. These are fungi that are common in most agricultural soils. Other fungi that infect vegetable seedlings are Fusarium species and Theilaviopsis basicola.

Wide Host Ranges

Many Pythium species have a very wide host ranges and can be found in most all agricultural soils. Various Pythium species can cause a seed rot and pre- and post-emergence damping off. Rotted seeds will be water soaked and mushy when squeezed. The roots of infected seedlings will be water soaked in appearance and gray in color. It may also cause a root rot in fields with poor drainage, with plants in wet areas being yellow and stunted.

Probably the most common species of Pythium infecting vegetables is P. ultimum, but there are many other species that have been associated with seedling diseases of vegetables. Pythium species generally survive as thick walled spores in the soil. Other types of spores can be produced, some which may swim to new areas on the plant root or to other nearby roots.

Rhizoctonia solani is also very common in most soils and also has a very wide range of plants that it can survive on. Symptoms of Rhizoctonia solani damping off are almost indistinguishable from plants infected by Pythium species. Symptoms include seed rot and damping off. The roots may be discolored, generally more so than with Pythium species.

Rhizoctonia solani survives in the soil as sclerotia on colonized organic matter from the previous crop. But it may also survive and reproduce on the roots of weed hosts. The survival of Rhizoctonia solani however depends on organic matter, either in plant debris or on living host such as weeds or a particular crop.

Biological Seed Treatments and Coatings

The purpose of any seed treatment is to improve seed performance in one or more of the following ways:

1. Eradicate seedborne pathogens or protect from soilborne pathogens,

2. Optimize ease of handling and accuracy of planting (reduce gaps in stand or the need for thinning of seedlings, particularly when mechanical planters are used), and

3. Improve germination rates. In conventional production, seed is often treated with chemical fungicides which reduce seed and seedling losses due to seedborne and soilborne disease. Most seed protectants are not an option for organic growers; however, there are some seed treatments, such as priming, pelletizing, and the use of hot water or NOP-compliant protectants, that can be used by organic farmers to improve seed performance.

Certain crops are better candidates for seed treatment due to the nature of the seed (small or irregularly shaped) or the intended production regime. For example, pelleted seed is useful in head lettuce production because of the need for precision seeding, but is less advantageous for thick sowings of looseleaf lettuce in bed production.

Priming

Primed seed has absorbed just enough water to dissolve germination inhibitors and activate the early stages of germination. Primed seed is therefore in a suspended state of growth, so it germinates faster and more uniformly over a broader temperature range, reducing the likelihood of very thick or thin plant stands. Priming results in earlier seedling establishment, which can aid in fending of the attack of damping-off pathogens to which germinating seedlings are particularly vulnerable. Priming is usually performed in conjunction with a pelleting process to protect the primed seed, which has a shortened life expectancy.

Pelleting

A seed pellet is a coating, usually of clay mixed with other inerts, that streamlines the size, shape, and uniformity of a small, non-round seed such as those of lettuce, carrots, onions, and many herbs and flowers. Pelleting results in easier, safer, and more accurate mechanical seeding, thus reducing gaps in the field and the need for labor-intensive thinning. Ideally, the pelleting materials are somewhat permeable to oxygen and absorb water quickly so that the pellet splits immediately upon hydration. Conventional pelleting techniques using synthetic inert materials are not approved for organic use, but there are now several pelleting materials on the market that are approved for use on organic farms.

Seed Health Treatments

This is a broad category of treatments that includes hot water, biological and plant extracts, bleach disinfection, and biologicals (microbes). These treatments can improve seed and seedling health by eradicating seedborne pathogens from the seed or protecting germinating seeds from attack by soilborne pathogens.

Hot Water Treatment

The use of hot water treatment to eradicate seedborne diseases, particularly those caused by plant pathogenic bacteria, is well-established. While the technique does not work for large-seeded vegetable crops, it has proven effective for brassicas, carrots, tomatoes, and peppers, and, to a lesser degree, celery, lettuce, and spinach. The typical procedure consists of: 1) warming the seed in 100 °F water, 2) heating the seed for 20-25 minutes, depending on the crop species, in a 122 °F water bath, 3) cooling the seed for 5 minutes in cold water, and 4) rapid drying. Precision in temperature and timing are important, as the seed embryo may be killed in hotter water or the disease incompletely eradicated in cooler water.

Crop	Temperature	Duration
broccoli	122 F	20 min
kale	122 F	20 min
mustards	122 F	20 min
collards	122 F	20 min
turnip	122 F	20 min
cabbage	122 F	25 min
cauliflower	122 F	20 min
Brussels sprouts	122 F	25 min
pepper[1]	122 F	25 min
tomato[2]	122 F	25 min
eggplant	122 F	25 min
carrot	122 F	20 min
celery	122 F	30 min
lettuce[3]	118 F	30 min

[1] pepper may be more sensitive than tomato to hot water trt
[2] can also try 125 F for 20 min
[3] lettuce is more sensitive; try small sample first and test viability

Hot water treatment can cause a reduction in vigor over time, so hot water treated seed should not be kept for longer than a season. Some companies do their own hot water treatment or will custom hot water treatment upon grower request. If a lot is not treated by the company and no testing has been done for pathogen detection, growers may conduct their own hot water treatment with a home set-up. It should be noted that the company's liabilities are null and void if the grower treats the seed him/herself. Only fresh seed of high vigor should be subjected to hot water treatment, as old seed or seed of low vigor may respond poorly to the stress of the treatment and have reduced viability. Hot water treated seed should be used within one season; the storage life of the seed may be reduced by the treatment.

Plant Extracts and Oils

Evaluating plant extracts and oils as seed treatments is a new research area so there is currently little data on their efficacy. However, plant oils such as thyme, cinnamon, clove, lemongrass, oregano, savory, and garlic show some potential to suppress damping-off, and thyme oil is in use in Europe as a seed treatment. Pure soybean or mineral oils have been shown to reduce storage molds of maize and soybean. Further research on the disease suppressive potential of these oils is necessary to determine the viability of essential oil-based seed treatment protocols.

Bleach Disinfection

Bleach (sodium hypochlorite) can be used to surface-disinfest seeds as an alternative to hot water. Bleach will eliminate pathogens on the seed surface but will not eliminate pathogens beneath the seed coat. Sodium hypochlorite is allowed for use on organic farms to disinfect wash water, provided that the levels not exceed the maximum residual contamination levels of the Safe Drinking Water Act, which currently is 4 ppm expressed as chlorine.

Biological Seed Treatments

Biological seed treatments, alone or in conjunction with priming and pelleting processes, may have potential in some situations for improving seedling health. In studies evaluating the efficacy of these microorganisms as seed treatments or drenches, results have been inconsistent.

Organic Seed Disease Treatment

Organic seed disease prevention starts with health-promoting cultural practices in ecologically managed farms that prevent disease and pests in the first place, that include:

- Crop rotation- creating disease-suppressive soil/compost, habitats for benefical pest predators.
- Appropriate Planting Dates, Soil Temperature and Moisture.
- Selecting Disease-Resistant Varieties.
- Cleaning and Processing Methods that Control Disease.

Seed Surface Treatments

Surface seed treatments reduce disease-causing fungi and bacteria found on the seed. Biological seed treatments control seed pests by parasitizing the pest organisms, competing for food on the root system, or producing toxic compounds that inhibit pathogen growth. Control of surface pathogens include beneficial microbes in compost teas, herbal sprays, washes or oils, hot water, heat, and disinfectants.

- Compost and Vermi-Compost Teas.

- Biodynamic Treatments.

- Herbal Treatments.

- Hot Water Bath.

- Disinfectants.

- Commercial Products.

- Indigenous Methods.

- Research and Resources.

Treatments to Prevent Soil-Borne Disease

This approach surrounds the germinating seed with protection from the disease-causing fungi, soil-borne pathogens that cause damping-off, and other soil-borne diseases. Fungal pathogens in the soil can rot the seed before it emerges from the soil or kill plants as they emerge. Treatments encourage healthy root systems.

- Drench Transplant with Compost Tea - inoculant to enhance beneficial soil microflora.

- Seed Surface Treatments.

Compost Tea and Vermi-Wash

Disease-suppressive compost has complex microbial communities that compete with and control pathogens. It is typically used to coat the leaf surface with beneficial microbes or as a soil drench. To produce suppressive compost, we suggest to pre-mix the raw ingredients, manage at lower temperatures on fertile soil, minimize turning and inoculate with earthworms to increase beneficial microbes. Vermicompost (earthworm castings) has been found to have significant disease-suppression. Research is needed to investigate use of compost teas for control of seed-surface pathogens.

Use of Specifically-prepared Compost Tea for Soil and Foliar Biocontrol.

Late blight of potato, tomato, Phytopthora infestans	Horse compost extract
Gray mold on beans, strawberries Botrytis cinerea	Cattle compost extract
Fusarium wilt Fusarium oxysporum	Bark-compost extract
Downy & Powdery mildew-grapes Plasmopara viticola, Uncinula necator	Animal manure-straw compost extract
Powdery mildew on cucumbers Sphaerotheca fuliginea	Animal manure-straw compost extract
Gray mold on tomato, pepper	Cattle & chicken manure compost extract
Apple scab Venturia conidia	Spent mushroom compost extract

Biodynamic

Seed Soaks and Sprays

Biodynamic preparations are used to enhance the biological activity of the soil. The preparations consist of mineral, plant, or animal manure extracts, usually fermented and applied in small proportions to compost, manures, the soil, or directly onto plants, after dilution and stirring procedures called dynamizations.

They are numbered 500 to 508. BD 500 is made from cow manure fermented in a bovine horn and buried for six months through autumn and winter. It is used as a soil spray to stimulate root growth and the production of humus. BD 501 is the horn silica preparation made from powdered quartz packed into a cow horn and buried in the soil through spring and summer for six months. It is used as a plant spray for the stimulation and regulation of growth. BD 502 - 507 are compost inoculates and are inserted into the compost piles to increase microbial activity, enhancing the decomposition process. They are made from the fermented herbs yarrow (Achillea millefolium), chamomile (Anthemis nobilis), stinging nettle (Urtica dioica), oak bark (Quercus alba), dandelion (Taraxacum officinale), and valerian (Valariana officinalis), respectively. Each preparation stimulates processes essential for plant growth and are used to strengthen the life forces on the farm.

The preparations are used a seed soak to encourage germination and growth of seedlings and for anti-fungal control. The solutions are lightly sprayed on seeds and then quick dried on a screen.

BD Seed Baths

Material	Seed	Instructions
Horn Manure	Spinach	Stir prep for 1 hour
Barrel Compost	Root crops	1 part BC + 4 parts rainwater + 5 parts milk, leave 24 hours, stir 5 minutes before use
Valerian	Beet, onion, tomato, potato	1 tablespoon/10 liters, stir 15 minutes
Yarrow	Grain, grasses	1 portion in 3 liters rain water, stir vigorously 5 minutes, leave 24 hours, stir before use
Chamomile	Legume radish, brassica	Same as for yarrow
Oak Bark	Oats, lettuce, potato, dahlia	Same as for yarrow
Nettle	Barley	Same as for yarrow

Herbal Treatments

- Equisitum, common name: Field Horsetail, is prepared as a tea to suppress fungal growth of mildew and other fungi, on grapevine, vegetables and trees. The useful variety, 'arvense' has a slightly discolored or green collar. The less potent varieties have a collar formed of small black pointed leaves around the joints. The plant propagates by spores that spread out from the brown leafless stems in spring. The plant is difficult to start but reproduces well once established. Prepare by boiling plants for a 20 minutes to a half an hour. Then spray on seeds.

Stinging Nettle (Urtica Dioica) is reputed to increase disease resistance in neighboring plants, possibly through a biochemical exudate and stimulates humus formation. Used as a nutrient rich spring tonic and used as a BD prep by fermentation. E. Pfeiffer sites an experiment showing that stinging nettle planted between rows of tomatoes decreased rotting and enhanced production of oils in pepperment plants. It is observed that nettles stimulate soil life decomposers to produce mature humus. Fermented nettles are added to BD compost.

Hot Water Seed Treatment

Kills most disease causing organisms on or within seed. Suggested for eggplant, pepper, tomato, cucumber, carrot, spinach, lettuce, celery, cabbage, turnip, radish, and other crucifers:

- Use carefully: Improper treatment can cause seed injury.

- Seed of cucurbits can be severely damaged by hot-water treatment.

- Pre-warm loose seed in woven cotton bag for 10 minutes at 100 F water. Place pre-warmed seed in water bath that will hold the recommended temperature.

Length of treatment must be exact. After treatment, dip bags in cold water to stop heating action, spread seeds to dry. Recommends applying protective seed tx fungicide to hot-water treated seed.

Old seed can be severely damaged by this treatment. A small sample of any seed lot over 1 year old should be treated and tested for germination to determine amount of injury that may occur. Thiram is most frequently suggested seed-protectant fungicide (do not use treated seed for food or feed).

Hot Water Treatment of Vegetable Seeds

- Cultural management: This is a broad category of management tactics that are numerous common-sense approaches that affect not only disease incidence and severity, but also plant growth and management of other pests, in both the greenhouse and field. The first consideration is to always use clean seeds, as free of pathogens as possible. For tomatoes, seeds should be tested for the bacterial pathogens causing bacterial canker, bacterial spot and bacterial speck. Certain fungi may also be seed borne, but testing is not routinely carried out for these pathogens. We recommend hot water treatment for all seed lots testing positive for a bacterial pathogen, and all untested seed lots. Hot water treatment is preferred because the bacterial canker pathogen survives inside the seed coat and is not completely eliminated by surface disinfestations using Clorox, acid, or other treatments. It is also permitted for organic tomato production. Tomato seeds should be placed in a loosely woven cotton (e.g. cheesecloth) bag, not more than half-full, and pre-warmed for 10 minutes in 100 degree Fahrenheit water. The bag should then be moved to a 122 degree Fahrenheit water bath for 25 minutes exactly. Old or poor quality seed may be damaged by this treatment.

Preventing Bacterial Diseases of Vegetables with Hot-Water Seed Treatment

Hot Water Treatment

Hot water treatments control many seed-borne diseases by using temperatures hot enough to kill the organism but not quite hot enough to kill the seed. Treatment for the fungal disease blackleg and the bacterial disease black rot of crucifers is a classic example of hot water treatment. It must be carefully and accurately done. Therefore, growers should ensure that heat treatment was performed by the seed company, avoiding performing hot water treatments themselves. A few degrees cooler or hotter than recommended may not control the disease or may kill the seed. The objective is to use as high a temperature and as long a duration as possible without seriously impacting germination. Hot water treatment can be damaging or not practical for seeds of peas, beans, cucumbers, lettuce, sweet corn, beets and some other crops. Some hybrid varieties of cauliflower may be damaged by the recommended treatment. Spores or bacteria that are attached to seeds can be killed by soaking the seeds in hot water. Use water of exactly 50 °Celcius and soak the seeds for 30 minutes.

Disinfectants

Clorox Treatment: removes bacterial pathogens on seed surface. Recommended for peppers, tomatoes, cucurbits and other vegetavles if the seed has not been treated by another method

To use for organic seed, this is not a treatment, but seed cleaning. Rinse with water after the washing with bleach.

Instructions: Agitate seeds in solution of 1 part Clorox in 4 parts water with one teaspoon surfactant for 1 minute. Use 1 gallon disinfectant solution per pound of seed (prepare a new solution with each batch) Rinse seed thoroughly in running tab water for 5 minutes, then spread and dry. Dust seeds with Thiram (75 WP 1 teaspoon/lb seed). Do treatment close to planting time.

Seed Soaked in Bleach or Acid

Soaking seed in a bleach or acid solution can be very helpful in control of certain bacterial pathogens that may be carried on the seed coat, including bacterial speck, canker, and spot or tomato or pepper. This is a treatment that growers can and should implement themselves. The time and concentration needed varies by crop.

Clorox Seed Treatment for Tomato and Pepper

Seed may be treated by washing 40 mins. w/continuous agitation in 1 part Clorox liquid bleach (5.25% sodium hypochlorite) to 4 parts water (i.e. 1 pint Clorox plus 4 pints water). Rinse seed in clean water immediately after removal from the Clorox solution and promptly dry. Germination may be compromised if washing time exceeds 40 mins.

Biological Seed Treatment For Corn and Soybeans

Bio-Seed-Gard is an OMRI-approved blend of micro-organisms for use as a seed treatment for Corn, Soybeans, Peas, Legumes, and other crops. AgriEnergy Resources, the Co. that developed the product, has done greenhouse and field trials showing improved: seeding vigor, stand establishment, root growth, plant growth, and yield. Dry blend of Trichoderma and Mycorrhizal species.

CB-QGG is a liquid biological seed treatment and root growth promoter formulated with beneficial microbes, macro and micro nutrients, amino acids, organic acids, root growth stimulants, enzymes, proteins, vitamins and minerals. CB-QGG makes less available forms of soil phosphate available to plants, promotes nitrogen fixation, root development and quick emergence, stimulates cell division and increases stress tolerance. In addition, it produces substances which increase the vigor of treated plants helping them resist damage by pathogenic fungi and damping off. This results in a more consistent plant stand and increased yields. CB-QGG promotes early root growth and larger roots. Larger roots provide better access to moisture and nutrients which translates into improved health throughout the life cycle of the crop.

When used as directed, the beneficial fungi in CB-QGG penetrate and colonize the plant roots and send out filaments into the surrounding soil. These filaments form a bridge that connects the plant roots with large areas of soil (up to 200 times larger than the root zone) and act as a "pipeline" to funnel nutrients to the plant. In return, the plant discharges compounds, through its roots, to stimulate fungal growth. The efficacy of the fungi is also enhanced by soil microbes such as bacillus sp. These bacteria, many of which are aerobic, affect root colonization and function by producing vitamins, hormones, and other compounds that promote fungal growth.

Biological Seed Treatment Fungal spores or bacteria can infect seeds while they are still developing on the plant or after the harvest. These pathogens cause diseases in the next crop (seed-borne diseases). Antagonistic fungi or bacteria can be used to protect seeds. Examples of these biological agents are Trichoderma sp. (antagonist fungus) and Bacillus subtilis (a bacterium). Biological seed treatments multiply in the soil, protecting root systems against soil-borne pathogens (ie: damping-off) after germination.

Trace Element Seed Treatment 'Application of trace elements to the seed provides nutrients to the seed during the critical period of plant establishment. The early alleviation of nutrient deficiency enhances crop yield and quality.'

Indigenous

Seed Treatment Methods

Seed Treatment 1: The cow urine treatment with 1:10 concentration was found very suitable to treat seeds of finger millet for good germination and seedling vigor. So farmers were advised accordingly to treat the finger millet with cow urine at 1:10 dilution.

Seed Treatment 2: Finger millet seeds are soaked with water for 18 hours (1 kg seed/600 ml water). The water is drained and seeds are dried under shade just 10 - 15 min before sowing. It helps in good germination and through this treatment it resists stress and overcomes drought condition very effectively.

Seed Treatment 3: Salt Water Treatment (Place an egg in water until it sinks to the bottom and rests horizontally. The horizontal egg indicates that the egg is not rot and that it could be further used for purpose of testing the density of salt concentration in water). Then remove the egg add salt to water and stir continuously until it dissolves. Immerse the egg once again into this concentration. This process of immersing the egg to test the density of salinity concentration continues. At a certain stage of salt application into water, only a quarter of the egg is found to be visible on the surface. This is the standard saline concentration used for seed treatment. Seeds are soaked in this preparation for a period of 30 minutes and then removed, washed twice and thoroughly dried until the moisture is absorbed. Salt-water treatment is a useful technique in separating healthy seeds from chaffy ones.

This treatment is found to be very effective in treating paddy seeds.

1. It separates the healthy seeds from chaffy seeds.

2. Protects from storage fungus.

3. Activates embryo.

References

*   Seedldiseas: tamu.edu, Retrieved 8 February, 2019

*   Seed-borne-pathogen-diseases-affecting-crops, introduction-transmission-and-detection-techniques, plant-diseases: biologydiscussion.com, Retrieved 17 May, 2019

*   Soybean-seed-and-seedling-diseases, pest-management: umn.edu, Retrieved 24 August, 2019

*   Cotton-seedling-diseases: ces.ncsu.edu, Retrieved 2 January, 2019

*   How-prevent-seedling-damping: umn.edu, Retrieved 25 June, 2019

*   Pastpest: ipm.illinois.edu, Retrieved 14 January, 2019

*   Variety-fungi-can-cause-seedling-diseases: farmprogress.com, Retrieved 19 April, 2019

*   Organic-seed-treatments-and-coatings: extension.org, Retrieved 29 July, 2019

*   Seedtreatments: growseed.org, Retrieved 9 March, 2019

Biocontrol of Rust Diseases and White Powdery Mildew

Rust refers to a number of plant diseases which are caused by pathogenic fungi. It commonly affects wheat and various ornamental crops like marigold and fuchsia. Powdered mildew is another fungal disease which affects a large number of crops such as wheat, barley, grapes and onions. The diverse applications of biocontrol in treating rust and powdered mildew have been thoroughly discussed in this chapter.

Rust Diseases

The rusts are a group of fungal diseases affecting the aerial parts of plants. Leaves are affected most commonly, but rust can also be found occasionally on stems and even flowers and fruit.

The spore pustules produced by rusts vary in colour, according to the rust species and the type of spore that it is producing. Some rusts have complex life-cycles, involving two different host plants and up to five types of spore.

The rusts are amongst the most common fungal diseases of garden plants. Trees, shrubs, herbaceous and bedding plants, grasses, bulbs, fruit and vegetables can all be affected. Rust diseases are unsightly and often (but not always) reduce plant vigour. In extreme cases, rust infection can even kill the plant.

Symptoms

You may see the following symptoms:

- Pale leaf spots eventually develop into spore-producing structures called pustules.

- The pustules are found most commonly on the lower leaf surface and produce huge numbers of microscopic spores.

- Pustules can be orange, yellow, brown, black or white. Some are a rusty brown colour, giving the disease its common name.

- In some cases there may be dozens of pustules on a single leaf.

- Severely affected leaves often turn yellow and fall prematurely.

- Pustules also sometimes form on leaf stalks (petioles), stems and, rarely, on flowers and fruit.

- Heavy infection often reduces the vigour of the plant. In extreme cases (e.g. with antirrhinum rust) the plant can be killed.

Rust Diseases of Wheat

There are three different rust diseases that affect wheat—leaf rust (also known as brown rust or orange rust), stripe rust (commonly known as yellow rust), and stem rust (commonly referred to as black rust of black stem rust).

The Rust Fungi

Leaf, stripe, and stem rust are caused by Puccinia recondita f. sp. tritici, Puccinia striiformis f. sp. tritici, and Puccinia graminis f. sp. tritici, respectively. These pathogens are specialized into numerous physiologic races that are identified by their reactions on an established set of differential wheat varieties. A complex system has been developed to keep track of the hundreds of known races. Any given variety may be immune, resistant or susceptible to a race of rust, but no variety is resistant to all races of any of the three rusts. Every few years new races of these fungi arise, causing previously resistant varieties to become infected and diseased. The life span of a rust resistant variety is usually from 2 to 4 years.

Symptoms

All rusts are typified by the presence of rusty-colored pustules erupting through the plant surface. They can be distinguished from other leaf diseases by rubbing or smearing the rust spores on the leaf surface with your finger. On wheat, leaf, stripe, and stem rusts are distinguished from each other based on the color, size, and arrangement of the pustules on the plant surface and the plant part typically affected.

Pustules of leaf rust, found predominantly on the leaf blade and sheath, are small, up to 1/16 inch long, round to oval fruiting bodies (uredinia) of the rust fungus. Reddish-orange urediospores develop within the uredinia and rupture the epidermis of the leaf surface as the spores mature. Pustules can be either scattered or clustered on the leaves and leaf sheaths of infected plants. Each pustule contains thousands of spores. Leaf rust developing from fall infections usually appears first on the lower leaves and progress up the plant to the upper leaves by mid-June. However, infections usually occur first on the upper leaves due to the fact that wind-blown spores are deposited out of the air during spore showers. Under severe epidemics, pustules may develop on the awns and glumes of the heads or occasionally on the stem below the head.

Strip rust pustules are yellowish-orange, much smaller than those of leaf rust, and are neatly arranged in groups forming distinct stripes on the leaf surface. For stem rust, on the other hand, pustules are much larger, orange-red, oval to elongated, and develop predominantly on the stem, leaf blade and sheath, and occasionally on parts of the spike. One of the most characteristic features of stem rust that helps to separate it from the other two rusts is the fact that the uredinia tear the plant tissue, giving the affects stem and leaf a distinctly tattered appearance.

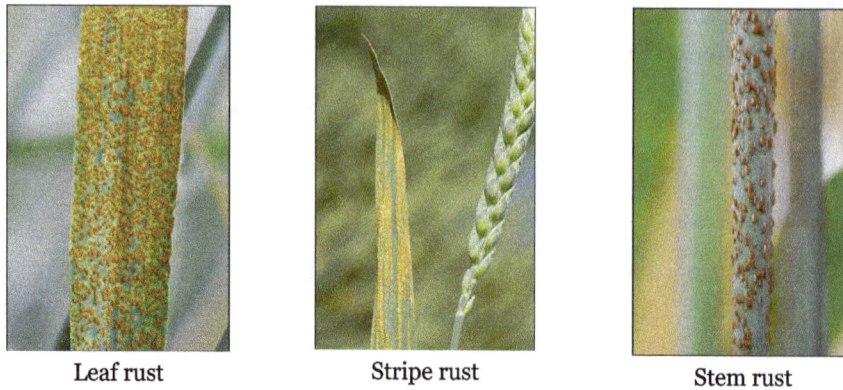

Leaf rust   Stripe rust   Stem rust

Disease Cycle and Epidemiology

Wheat rusts have very complex life cycles that include two hosts (wheat, the primary host, and an alternate host) and five different spore stages. However, the in-season rust cycle in Ohio is fairly simple, since the alternate host and three of the five spore stages are of little importance for in-season rust development. In Ohio and other parts of the Midwest, the urediospore stage is the spore type responsible for dispersal and infection of the wheat crop. Urediospores overwinter on infected wheat in the more moderate climate of the southern states and Mexico, and are carried northward by the wind. Under favorable temperature and moisture conditions, urediospores germinate and infect leaves within 6 to 8 hours after landing on the plant surface. Once established, a new crop of urediospores may be produced every 7 to 14 days if environmental conditions are favorable. The earlier rust develops, the more spore and disease cycles are likely to occur during the season and the greater the risk of severe epidemics and yield loss. Frequent heavy dew, light rain, or high humidity and temperatures of 77 to 86 °F are ideal for leaf rust development. Stem rust has a similar optimum temperature range, but stripe rust develops best under much cooler conditions (50 to 64 °F). All three diseases are spread by wind blow urediospores from plant to plant and from field to field until the crop matures. As the plant matures, black, submerged pustules develop on the leaves, leaf sheaths, stems, and spikes, depending on the rust. These pustules (telia) contain the winter spores (teliospores). Teliospores do not infect wheat. Telia may not develop when plants become infected very late in the season (close to maturity). In the fall urediospores are blown southward and infect wheat and overwinter as urediospores or mycelium on volunteer wheat plants.

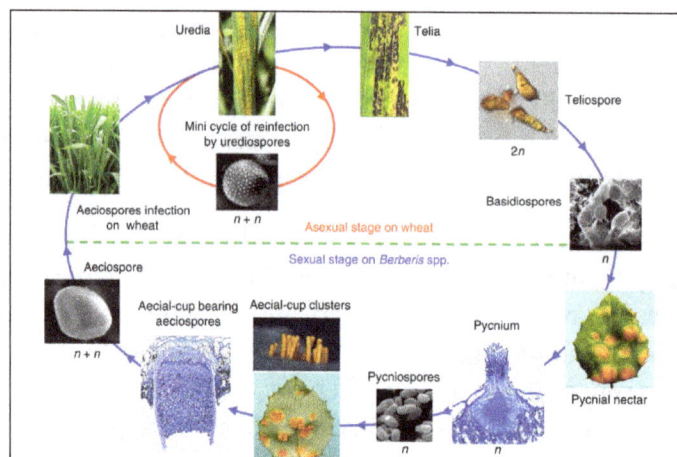

Rust Diseases of Ornamental Crops

Rust diseases are common fungal infections that affect a wide range of floricultural crops, including Aster, Carnation, Fuchsia, Florist's geranium (Pelargonium X hortorum) Gladiolus, Lilium, Marigold, Poinsettia, Snapdragons, Statice and Viola (including pansy). Rusts have the potential to negatively impact floriculture production because these pathogens often cannot be detected on infected, but symptomless propagation material entering the United States or moving state-to-state. Rust fungi are obligate parasites, dependent upon a live host for growth and development, and seldom kill plants. However, rust infection reduces plant health and vigor, flower production, and aesthetic value.

Symptoms

Each type of Rust has its own distinctive symptoms and its own specific plant hosts. The disease often first appears as chlorosis on the upper surfaces of leaves. All rust fungi produce powdery masses of spores in pustules, typically on leaf undersides that are yellow, orange, purple, black or brown. Some Rust fungi produce pustules on upper leaf surfaces as well. Spores are easily spread on air or with splashing water. Lesions may coalesce resulting in large areas of necrosis; leaf distortion and defoliation often follow.

Aster Rust

Aster is affected by several rust diseases-Coleosporium campanulae, Puccinia asteris, P. campanulae, and other Puccinia species. Orange-red pustules develop on the leaf undersides; heavy infections can cause leaf yellowing and necrosis. C. campanulae requires pine as an alternate host, while Puccinia species have various sedges and grasses as alternate hosts. The removal of alternate hosts is a good management strategy.

Carnation Rust

Caused by Uromyces diantha, this rust disease is characterized by small pustules of powdery, brown urediospores. These spores are carried over only on live plants. Resistant cultivars are available.

Chrysanthemum Rust

Two species of Puccinia causes rust on chrysanthemums P. chrysanthemi and P. horiana. P. chrysanthemi is most common in late summer and is characterized by dirty-brown pustules and yellowish-green spots on upper surfaces of leaves. P. chrysanthemi causes minor damage in the field and is uncommon on greenhouse plants. Severe infestation may damage large areas of leaves and lead to defoliation and reduced flower production. Chrysanthemum varieties resistant to rust include 'Achievement', 'Copper Bowl', 'Escapade', 'Helen Castle', ' Mandalay ', 'Matador', 'Miss Atlanta', 'Orange Bowl', and 'Powder Puff'. P. horiana causes Chrysanthemum white rust and as a recent introduction to the United States is subject to quarantine and an eradication program. Symptoms are white, pinkish or brownish pustules produced on leaf undersides with white, yellow, to pale-green lesions on upper leaf surfaces. Chrysanthemum white rust results in leaf distortion, discoloration, defoliation, and plant death. White rust is primarily a disease of greenhouse crops; when it occurs outside direct sunlight and low humidity kill the spores. Contact state and federal agricultural officials if any suspect white rust infections occur and destroy all plants. Regulations require that infect plants be destroyed to prevent disease establishment in this country.

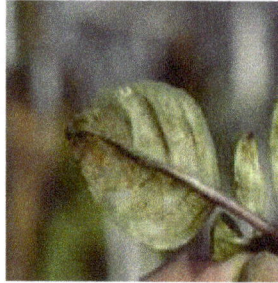

Fuchsia Rust

Fuchsia rust is caused by the fungus Pucciniastrum epilobii and occurs throughout the United States. The most serious losses occur during propagation; diseased plants at any stage are not marketable. Affected leaves may be deformed and defoliation often occurs. Weakened plants may occasionally die. The initial source of spores may be from the alternate hosts, fir or fireweed, especially if plants are kept outside in the summer, or from infected stock brought in from other sources.

Geranium Rust

Geranium rust caused by Puccinia pelargoni-zonalis is most serious on Florist's geraniums (Pelargonium X hortorum), but has also been reported on zonal geraniums and seedling geraniums. Ivy geranium (P. peltatum), Martha Washington or regal (P. X domesticum), the scented leaf types, and the wild geraniums are resistant. Geranium rust occurs throughout the United States due to the ease with which it is spread on infected cuttings. This rust spends its entire life cycle on geranium (autoecious). Symptoms first appear as small, circular, yellow spots on the top of leaves opposite the pustules on the lower leaf surface. The spots on the lower leaves enlarge to blister-like pustules of rust to cinnamon brown spores which often develop in concentric rings. The purchase of certified, culture-indexed cuttings from a reputable commercial propagator is an important means of disease control.

Poinsettia Rust

Unlike most other rust fungi, pustules of Poinsettia Rust caused by Uromyces euphorbiae occur on both leaf surfaces. Spores are cinnamon brown and occur only on living plants.

Rose (Rosa Species) Rust

Rose rust caused by Phragimidium mucronatum first appears in the spring as bright orange pustules on leaf undersides, leaf stalks and branches. In the summer, small raised orange spots appear on the undersides of leaves with small yellow specks on the upper surface beginning on the lower leaves and moving upward. The spots may go unnoticed until the plant begins to exhibit a generally unhealthy appearance and a loss of the lower leaves. Leaves may become dry and twisted before falling off. As the leaves dies, the spots darken and produce the black teliospores that survive the winter to produce new infections in the spring. This disease may be mild or severe depending on the season and how early the infection began. Prune and burn infected leaves, petals, and canes as soon as they are detected. Cultivars vary in their susceptibility to rust diseases.

Viola species Rust

Caused by Puccinia viola, this rust disease is common in the Northeast. Symptoms first appear as small, pale green spots on the upper leaf surface. As the fungus develops within the leaf, corky spots, blisters, or pustules containing rusty brown spores develop on leaf undersides.

Biocontrol of Rust Fungi by Cladosporium Tenuissimum

The use of chemical pesticides in agriculture is under increasing public scrutiny, with mounting concern over possible harmful effects on the environment and on human health. Agrochemicals pollute groundwater, enter the food chain, have a deleterious impact on many living organisms and may bring about pesticide resistance in plant parasites. Increasing awareness of such possible hazards is prompting a heightened interest in alternative strategies.

One of the most promising approaches to control economically important crop pests and diseases is by exploiting naturally occurring antagonists. Over the past few years unprecedented advances have been made in the biological control of many damaging insect pests and phytopathogenic microorganisms.

Many fungal pathogens throughout the world have natural enemies that limit the harm they cause. Some of these competitors are non-fungal hyperparasites such as bacteria or mycoviruses, but most are other fungi. Antagonistic fungi are a major component of the microbial community on the plant rhizosphere and phyllosphere, and play a pivotal role in regulating many interactions between plants and parasitic microorganisms. Since they are an integral part of the ecosystem, no alien microbe species, toxic substances or chemicals are introduced into the environment by their use, and hence they appear to be more environmentally sustainable than other, more intrusive, control methods. As a result, fungal biocontrol agents (BCAs) are becoming a promising means to control fungal diseases and to reduce dependence on chemical pesticides.

However, developing a reliable and effective method for the biocontrol of a fungal pathogen is not a straightforward process. Several biological control experiments that were successful in vitro produced inconsistent results in the field. Moving from the laboratory or the greenhouse to large-scale field-testing is difficult because, in the field, the antagonist is subject to environmental influences. Plant disease is the result of a dynamic interaction over time between a pathogen, a plant and the environ- ment. The environmental component of the disease triad is crucial for the success of a BCA, the limited ecological amplitude of which might represent a major constraint. While an antagonist has to actively control a pathogen, a period that may take from a week to several months, it must be able to withstand fluctuations in the physical environment and resist the competing action of other, established microorganisms. The insufficient ecological fitness of the antagonist inoculum may lower its effectiveness as a control agent, while a poor shelf-life may cause lack of persistence, which means that the control achieved does not last over a sufficiently long term.

Cladosporium Species Hyperparasitic on Rusts

The genus Cladosporium is one of the most widespread and prevalent genera of fungi, containing over 500 species Some of its species have teleomorphs in the ascomycete Mycosphaerella (Dothideales), but a vast majority of taxa are known by their anamorphic state or by criteria such as host association or presumed host specificity.

These additional criteria are used to support the traditional classification because the inadequacy or lack of description for several members, and the phenotypic plasticity within individual taxonomic entities, make typification of species difficult or impossible. Substrate differences, climate and geographic variations also influence morphological expression. As a consequence, morphological features such as conidium size, shape, septation, pigmentation, surface characteristics (smooth or with ornamentation), conidiophore size and morphology are often variable and inconsistent.

Members of this genus display a variety of lifestyles: some commonly occur on their hosts as epiphytes, endophytes, or as pathogens; others thrive as saprophytes, or even as hyperparasites.

Several taxa in this large genus are prevalently associated with rust sori, and some are assumed to be invariably hyperparasites of Uredinales. Cladosporium uredinicola Speg. is a common necrotrophic hyperparasite that destroys rust hyphae and causes coagulation and disintegration of the cell cytoplasm of a number of hosts, such as Puccinia cestri Dietel and P. Henn., Puccinia recondite Roberge ex Desm., Cronartium quercuum (Berk.) Miyabe ex Shirai f. sp. fusiforme, Puccinia violae (Schum.) DC and Puccinia horiana. Spegazzini reported that Cladosporium uredinophilum colonized and destroyed Uredo cyclotrauma Speg. propagules in Paraguay. Steyaert described Cladosporium hemileiae Steyaert as an effective hyperparasite of the coffee rust fungus Hemileia vastatrix in Zaire (Democratic Republic of Congo). Powell encountered Cladosporium gallicola in galls of Cronartium comandrae Pk. on Pinus contorta var. latifolia and considered the fungus parasitic on aeciospores and responsible for lowering aeciospore production. Sutton reported a close association between C. gallicola and Endocronartium harknessii (J.P. Moore) Hiratsuka on Pinus banksiana, and observed hyphae of the hyperparasite penetrating the rust aeciospores. Tsuneda and Hiratsuka found that C. gallicola parasitized E. harknessii both by simple contact, disintegrating the cell walls of the spores, and by actual penetration of the spores walls, with or without the formation of appressoria, and causing the coagulation and disappearance of the host cytoplasm. These authors also found evidence that the rust spores, acting by contact, secreted enzymes in order to disintegrate the cell walls. A similar behavior was noted for Cladosporium aecidiicola, a fairly common hyperparasite of rusts in Europe and in the Mediterranean area, parasitizing E. harknessii on Pinus contorta, Pinus muricata and Pinus radiata in California. This hyperparasite also heavily parasitized aecia of Puccinia conspicua (Arth.) Mains. in Arizona and urediniospores of Melampsora medusae Thüm under storage conditions. Srivastava et al. found that Puccinia horiana was regularly parasitized by Cladosporium sphaerospermum. Sharma and Heather reported that C. tenuissimum was a common phylloplane fungus that actively colonized urediniospores of the rust Melampsora larici–populina Kleb. on Populus × euroamericana. More recently C. tenuissimum was also detected in collections of aeciospores of the two-needle pine stem rust Cronartium flaccidum and its non-host alternating form Peridermium pini.

This list of Cladosporium species parasitic on rusts is not exhaustive and is likely to increase in the near future. The lack of genuine morphological structures and the great instability of such characters not only hamper the placement of Cladosporium species in a congruent evolutionary context, but also make it impossible to identify consistently many individual taxa and distinguish them from related ones. As new molecular techniques become increasingly employed to confirm and refine known taxonomic relationships, it is predictable that new taxa now undistinguishable or considered synonyms, will be added to the list. The number of hyperparasites is also expected to grow once it is shown that some Cladosporium species, now thought to be common saprophytes because they are found on aged host structures, are in fact true hyperparasites.

Table: Cladosporium species hyperparasitic on rust fungi.

Parasite	Host
C. aecidiicola Thüm	Endocronartium harknessii; Melampsora medusae Thüm; Puccinia conspicua(Arth.) Mains.,
C. gallicola Sutton	Endocronartium harknessii; Cronartium comandrae Pk.
C. hemileiae Steyaert	Hemileia vastatrix Berk. et Br.
C. sphaerospermum Penzig	Puccinia horianaHenn.
C. tenuissimum Cooke	Melampsora larici –populina; Cronartium flaccidum G. Winter; Peridermium pini (Pers.) Lév.
C. uredinicola Speg.	Puccinia cestri Dietel and P. Henn.; Puccinia recondita Roberge ex Desm; Cronartium quercuum (Berk.) Miyabe ex Shirai f. sp. fusiforme: Puccinia violae (Schum.) DC; Puccinia horiana
C. uredinophilum Speg.	Uredo cyclotrauma Speg.

Identification and Distribution of Cladosporium Tenuissimum

The dematiaceous hyphomycete Cladosporium tenuissimum has long been known as a polyphagous saprophyte occurring in the air, on the soil and on plant surfaces. It is a frequent colonizer of senescent or dead plant material and is common on the phyllosphere and rhizo- sphere of plant species, but it is also reported as an endophyte and a facultative plant patho- gen, causing blights, leaf spots, and seed, fruit and blossom-end rots. Together with other Moniliales, it is the aetiological agent of human mycoses known as chromoblastomycosis, phaeo- hyphomycosis and eumycotic my- cetoma. The hyper- parasitic nature of C. tenuissimum was first reported, as already mentioned, by Sharma and Heather on the poplar rust M. larici–populina in Australia. It has recently attracted the attention of research- ers because it suppresses rust under laboratory conditions and in glasshous- es, and because it secretes compounds that are biologically active against various phytopathogens.

The culture characteristics of this mitosporic fungus (colony appearance, texture and morphology, growth/temperature relationships, etc.), its micro-morphology (type of conidia and conidiogen- esis) and other distinguishing features, are summarized in Moricca et al. The effect of nutrient composition on the production of secondary metabolites has also been investigated in vitro on different agar media and liquid cultures. Irrespective of the culture medium, the fungus produces typical, geniculate and sympodially elongated conidiophores, by virtue of which it is unambigu- ously ascribed to the anamorph genus Cladosporium Link.

Since the abundance of heterogeneous taxonomic elements makes traditional classification at spe- cies level difficult, representative European isolates of C. tenuissimum were identified by matching mycological characteristics with nucleotide sequences from coding and non-coding regions of the ribosomal RNA operon and by chemotaxonomic profiling. Molecular differences in nucleic acid sequences were instrumental particularly in vis-à-vis species recognition, since congeneric, mor- phologically similar taxa (C. herbarum, C. cladosporioides and other unidentified Cladosporia) were syntopic to C. tenuissimum, often sharing the same ecological niche (rust sori).

Several Cladosporium members have not yet been comprehensively described, many of its taxa are still ill-defined and the appropriateness of their separate recognition remains a vexed question. Many species are not considered good species because they have not been identified with sufficient confi- dence. As a result, a number of ecological, phytopathological and biomedical studies identify Clad- osporia only at the generic level. This means that information on the geographic distribution of C. tenuissimum in particular is scanty. Direct and indirect elements suggest, however, that it is an ubiq- uitous fungus with a worldwide distribution. It is known to be cosmopolitan and has been recovered from various matrices. In Europe it has been found on rust aeciospore samples of two-needled pines from various countries. On the other hand, a high incidence of airborne Cladosporium inoculum was reported in the London region by Ainsworth. More importantly, that author also found that Cladospo- rium inoculum reached a peak in summer, coinciding with maturation of most rust spore stages. The possibility for the hyperparasite to colonize spores and control their numbers at developmentally just the right time is particularly attractive in the light of its potential exploitation for rust biocontrol.

The Hyperparasitic Relationship

The great potential of C. tenuissimum as a BCA has been evident since the initial experiments of Heather and Sharma and Sharma and Heather. These authors observed both direct parasitism of urediniospores of M. larici–populina by this hyperparasite, and inhibition of rust spores without any physical contact, suggesting that antibiosis might also occur. This evidence induced these re- searchers to postulate that C. tenuissimum had significantly reduced the incidence and severity of rust disease in poplar plantations in the Canberra district for several years.

The antagonistic capability of C. tenuissimum was further confirmed when it was found to inhibit, in vitro, the germination of propagules of other rust fungi. Selected isolates of C. tenuissimum significantly reduced average percentage germination of aeciospores of C. flaccidum and P. pini at 12, 18 and 24 h from inoculation with conidial suspensions of the hyperparasite (33, 39 and 46% versus controls, respectively). The germination of urediniospores of the bean rust fungus Uromyces appendiculatus treated simultaneously with a conidial suspension of the antagonist isolate 'Itt21' was reduced significantly from 56.4% to 36.9% after just 3 h of contact, a reduction of 35%. After 6 h, urediniospore germination was still lower than in the control (69.5 versus 83.3%), and at the end of the observations the difference was 17%. Beyond this time, U. appendiculatus urediniospores no longer germinated. Freshly collected aeciospores of the pine twist rust Melampsora pinitorqua and of the common rust Puccinia sorghi inoculated with a mixture of conidia from different antagonist isolates displayed, after 24 h, reductions in germination of 19% and 21%, respectively, compared with the controls.

Timing of Infection

The reduction in spore germination might also depend on the time of initial infection. Experiments on C. flaccidum and P. pini indicated that the order in which the rust and hyperparasite were deposited was the main factor causing variability in aeciospore germination. In these experiments three deposition sequences were tested:

1. Aeciospores deposited 1 h prior to the hyperparasite conidia;

2. Conidia deposited 1 h prior to the aeciospores;

3. Aeciospores and conidia deposited simultaneously. Maximum inhibition of aeciospore germination was achieved when conidia were inoculated 1 h before the rust aeciospores.

This outcome suggests that, in nature, control is most effective if the antagonist establishes itself early on the host surfaces, before the rust, in order to build up a mass of inoculum sufficient to parasitize rust propagules as they burst from the plant epidermis somewhat later. However, the research data do not present a coherent picture. Other studies on Puccinia recondita, Cronartium fusiforme and Cronartium strobelinum found that with these rusts, infections became most severe when the antagonist Darluca filum was inoculated before them.

Interactions in Storage

The recovery of C. tenuissimum at various latitudes, from separated geographic areas with varying climates, suggests that the fungus can survive and remain active under disparate environmental conditions. This adaptability is also shown by its ability to survive and parasitize aeciospores of C. flaccidum and P. pini in storage in a range of temperatures. In tests it was effective at −20, 4 and 20 °C. Control was greatest at 20 °C but spore viability was decreased at all test temperatures, including −20 °C. Viability also gradually decreased with storage time. Antagonisttreated spore lots were visibly discoloured. Stereoscope observations revealed deterioration of the rust propagules, which were densely intertwined with hyphae of the antagonist. The hyperparasite had proliferated extensively, giving rise to an appreciable mycelial biomass. Spore deterioration was the first indication that an exogenous enzymatic effect was probably involved in the disintegration process. The long shelflife of

C. tenuissimum inoculum at low temperatures provides evidence of ecological tolerance, a characteristic that is common to other hyperparasites. As in other fungal host–hyperparasite interactions, the antagonist thrives on the host spores, which represent an ideal pabulum However, the host propagules are not vital for the antagonist which, being also a facultative saprophyte, a plant and animal parasite and an unspecialized hyperparasite, can survive on various materials such as plant debris, foliar exudates, small insects and other organic substances occurring in the environment.

Rust spores collected during adverse weather (high humidity, rainfall) and not properly dehydrated before storage, soon clumped together, and were impaired and discoloured. Many of these clumped spores were soon heavily overgrown with hyperparasite mycelium, which proliferated and sporulated profusely on them, markedly decreasing spore viability. High humidity is therefore favourable to infection, pathogenesis and sporulation of C. tenuissimum.

Effect of C. Tenuissimum Culture Filtrates

Treatment of all rust spores with culture filtrates showed that enzyme(s) and/or toxic agent(s) had a role in C. tenuissimum parasitism. Aliquots (25 ml) of culture filtrates from four antagonist isolates spread separately on sterilized water-agar slides, dusted with aeciospores of two-needled pine rust fungi and incubated in the dark at 22 °C in moist Petri dishes, strongly reduced spore germination after 12, 18 and 24 h, as compared with the controls. The rust provenances, the antagonist isolates and the interaction term (rust provenance × antagonist isolate) were not significant variables, indicating an absence of physiological specialization in the hyperparasite. Several of the spores examined individually under the microscope for viability were barely recognizable, with their spinules displaced and scattered all over the mounting medium, indicating an action of proteolytic enzyme(s) secreted into the medium.

The pronounced sterility of rust spores treated with C. tenuissimum culture filtrates suggested that the hyperparasite was a source of extracellular antifungal antibiotics, as it is in other hyperparasite–parasite interactions. The toxicity of the culture filtrates was immediate, from the first inspection (after 12 h), and remained fairly constant for the duration of the experiment, with a slight increase in germination after 18 and 24 h.

Hyperparasitic Secondary Metabolites

A band of killers C. tenuissimum actively produces metabolites with antifungal properties. These include a major pure common metabolite with the molecular formula $C_{20}H_{16}O_6$, and corresponding to Mr 352, and a series of related compounds, all of which were isolated from the ethylacetate crude extracts (EtOAc CEs) of several antagonist isolates. This metabolite was already known as cladosporol, a dimeric decaketide which had been isolated from C. cladosporioides. Cladosporol induces hyphal malformations in Phytophthora capsici when tested at 10 mg/disc. It is also an inhibitor of b-1,3-glucan synthetase, the enzyme that synthesizes the fungal cell wall component b-1,3-glucan. In an in vitro assay with labelled UDP-glucose and b-1,3-glucan synthetase prepared from Saccharomyces cerevisiae, cladosporol showed an IC50 activity on the enzyme at 50 mg/ml.

The series of related compounds purified from EtOAc CEs of C. tenuissimum cultures, consists of cladosporols B, C, D and E. The major metabolite, now named cladosporol A, was isolated as a white powder and represented more than 30% of the crude extract. A second metabolite had the same 1H and 13C nuclear magnetic resonance (NMR) spectra as cladosporol A, except that it had a 4-oxo function instead of the C(4)HOH grouping. This metabolite was named cladosporol B. A third compound had the same basic skeleton as cladosporol A, but with a C(2)H2-C(3) H2 unit instead of the 2,3-oxirane ring. This compound was compatible with the molecular formula C20H18O5 and was named cladosporol C. A fourth metabolite, cladosporol D, was a cream-coloured solid with the formula C20H18O6. Its 1H and 13C NMR data, when compared with those of cladosporol C, indicated that it contained a C(3)HOH fragment instead of CH2, the remaining signals being quite similar. The last metabolite, cladosporol E, was isolated as a brown solid with the formula C20H18O7. Its 1H and 13C NMR spectra were very similar to those of cladosporol D, the only important difference being that this cladosporol had an additional hydroxy group at C-2.

Cladosporols A–C, produced in an amount sufficient to enable some assays on their biological activity, inhibited, in vitro, a number of rust fungi, non-rust fungi, Oomycota and yeasts. They suppressed germination of urediniospores of U. appendiculatus and of aeciospores of M. pinitorqua, C. flaccidum, P. recondita and P. sorghi, in a range between 75 and 100%, when tested at 100 mg ml^{-1}. Cladosporol B was the most active of the group, completely suppressing germination of U. appendiculatus at 50 ppm, reducing it by more than 90% at 25 ppm, and lowering it even at 12.5 ppm. Cladosporol A, through less inhibitory than cladosporol B, was more active than cladosporol C, reaching an inhibition value higher than 80% at the highest concentration.

The cladosporols reduced radial growth of colonies of the phytopathogenic fungi Alternaria alternata, Botrytis cinerea, Cercospora bieticola, Cercosporella herpotrichoides, Colletotrichum lindemuthianum, Fusarium roseum, Helminthosporium oryzae, Mucor sp., Rhizoctonia solani, Septoria tritici; of the Oomycota Phytophthora capsici, P. cinnamomi, P. erytroseptica, P. nicotianae, Pythium ultimum; and of human-pathogenic strains of Candida sp. The strongest antagonistic effect was against the Oomycota.

Sensitivity of tested fungi varied in relation to concentration and differences in the functional groups bound in this family of metabolites to the tetralone skeleton. The antifungal activity of described cladosporols is likely to reside, as reported for other similar compounds, in the intrinsic toxicity of the 2-tetralone chromophore and the occurrence of highly reactive substituents, like the epoxy group b-1,3-glucan, as a constituent of the fungal cell wall skeleton. By inhibiting the enzyme that synthesizes this constituent, the cladosporols directly affect the biochemistry and structural organization of the fungal cell. They thus strongly condition the pathogenicity of C. tenuissimum and play a major role in its hyperparasitism.

In Planta Assays

Inoculation tests on whole rust-infected plants in a controlled environment (greenhouse or laboratory) give a first indication of how the rust may be controlled in nature. They therefore represent an important step in studying the mode of action of the hyperparasite. If a natural inoculation procedure and an objective disease evaluation protocol are followed, in planta assays can, in just a few weeks, provide valuable data on the biocontrol effectiveness of a tested microorganism. Furthermore, in such an artificial system, the tri-trophic interaction between host plant, rust parasite and hyperparasite can be more accurately investigated, since the effect of the environment, which in the field can positively or negatively affect each interacting partner, is eliminated.

In two consecutive growing seasons, 1999 and 2000, spermogonia of C. flaccidum that had developed on 2-year-old, rustinfected pine seedlings, were inoculated with mixed conidial suspensions from different C. tenuissimum isolates. Rustinfected control seedlings were sprayed only with sterile water/Tween 20. Disease evaluation, based on a standardized procedure, was completed after 5 months, and the incidence and severity of the rust infection were defined. The percentage of infected seedlings and the number of infections per seedling stem were significantly lower in the antagonist-treated seedlings than in the untreated controls. Percentage mortality was significantly lower in seedlings with antagonist inoculation than in those without. A mycelial biomass attributable to the hyperparasite and detectable as a felty, dark greenish-brown mycelium was observed on spermatial and aecial fructifications on the bark of some treated seedlings. The fungus was positively identified by microscope examination of the sporulating structures (erect, straight, regularly septate conidiophores; holoblastic, conidiogenous cells; cylindrical to clavate ramo-conidia with 2–3 flattened, thickened scars; intercalary and terminal conidia of various shapes and sizes).

In an experiment exploring the hyperparasitism of C. tenuissimum on U. appendiculatus, a classical disease escape mechanism may have prevented the antagonist from parasitizing the rust. Primary leaves of bean plants of P. vulgaris L. cv. 'Borlotto nano Lingua di fuoco' were inoculated on their lower surface with a suspension of U. appendiculatus urediniospore and simultaneously treated with a suspension of C. tenuissimum conidia, or with a MPGG (malt extract, peptone, glucose, glycerol) culture filtrate of a 2-week-old liquid stationary culture of the hyperparasite. Controls were healthy bean plants inoculated with:

1. A water suspension of C. tenuissimum conidia;

2. Rust urediniospores, then treated with sterile water/Tween 20; or

3. Only a sterile uninoculated MPGGbroth.

Disease severity was scored by the number of pustules per square centimetre in a total of eight 1-cm² leaf areas on digitalized primary leaves, after 13 d from inoculation. After 1 month, plants inoculated simultaneously with U. appendiculatus and the conidia suspension developed rust in the normal way. By contrast, treatment with the antagonist culture filtrate provided total protection: the urediniospores did not germinate and the bean plants did not develop any infection.

Apossible explanation of these findings on bean rust is that in the simultaneous infections the time available for inter-fungus interaction was too short to enable the antagonist to establish a parasitic relationship with the pathogen. The specialized, biotrophic agent found refuge by rapidly penetrating into the living host tissues, occupying the ecological niche it has evolved since primordial times to colonize in order to gain access to nutrients and protection from natural enemies. The antifungal compounds and enzymes in the culture filtrate, on the other hand, acted immediately and prevented propagule germination and rust development.

Detached Leaves

A simple test in a strictly controlled environment, using Petri dishes and a water saturated atmosphere, can give an insight into the type of hyperparasitic interaction. Primary leaves of bean rust-infected plants, inoculated as above with a conidial suspension or a culture filtrate of C. tenuissimum were immediately detached and incubatedin 15-cm diameter Petri dishes. The leaf stems were dipped in a medium containing 1% water-agar (WA) supplemented with 5 ppm gibberellic acid (GA). Hyperparasite colonization of rust pustules and disease severity were monitored daily under a stereoscope, starting 1 week after hyperparasite inoculation and continuing until the end of the experiment. Rust regularly formed appressoria at the precise location of the stomata (indirect-type, dikaryotic penetration) but these appressoria began to collapse a few hours after formation. In spite of the positive tropism of C. tenuissimum conidia towards the rust propagules, as indicated by the many conidia closely attached to rust spores and appressoria, penetration of the bean leaves by the rust could not be prevented completely, and this can explain why, compared with the controls, disease severity was curtailed by 13% only. As in the experiment with the whole plants, the culture filtrate had a toxic effect on the detached leaves that prevented any rust spores from germinating.

LM, SEM and TEM Examination of the Host/Parasite Interface

Examination of the interface between C. tenuissimum and the rust agents C. flaccidum, P. pini and U. appendiculatus with light (LM), scanning (SEM) and transmission electron microscopy (TEM) showed the strong antagonistic action of the hyperparasite, and the multiple strategies it employed to parasitize the rust fungi. The reproductive capacity of C. tenuissimum appeared greatly enhanced by the proximity of rust spores, most of which were inactivated and overgrown by the antagonist mycelium. The hyperparasite sporulated profusely on the spores, producing on their surface tufts of fructifications bearing numbers of conidiophores. These conidiophores generated a multitude of asexual propagation units, represented by ellipsoidal to limoniform ramo-conidia, oblong or fusiform intercalary conidia, and mostly globose or subglobose terminal conidia. Prolific antagonist sporulation gave rise to an enormous mass of conidia which, together with the germ tubes and hyphae growing from them, were attracted to the rust propagules, to which they became firmly attached.

The contact stimulus between the host and the hyperparasite – a reflection of recognition events among them – mediated the secretion of different substances at the host–parasite interface. These substances are believed to play a fundamental role in pathogenesis, either in interactions between plants and parasitic fungi or between those fungi and their hyperparasites. Some of these substances, visible as a dense network of amorphous, fibrous material, were the direct product of hyperparasite metabolism and served to ensure close adhesion of the hyperparasite to the host cell wall. Amorphous material from a different source was observed adjacent to shrunken and eroded parts of the spore wall. This second type of extracellular, amorphous material characteristically accompanied spore penetration and was associated with the hyperparasite structures involved in the process, i.e. the variously shaped appressoria formed on the host surface and the infection hyphae. Ultrastructural examination at contact points provided evidence that lytic enzymes caused degradation of the host cell walls and released the amorphous material at the rust–hyperparasite interface. A decrease in electron density from the outer to the inner layers of the spore wall sometimes made the fibrillar chitin structure of the wall visible.

An adhesive matrix pad intimately connected the host and the hyperparasite to each other. This pad serves a double function, to attach and support the appressorium and the penetrating hypha, and to be a reservoir for enzymatic penetration. Such functions have already been reported both in inter-fungus parasitism and in host plant– parasite interactions. The growth of the antagonist on a synthetic medium containing the polymer laminarin as the sole carbon source, on the other hand, shows that it produces extracellular b-1,3- glucanases. There is extensive enzymatic degradation of the matrix wall polysaccharides in which the polymeric chitin microfibrils in rust fungi are embedded, especially in the early stages of rust–hyperparasite interaction, when nutrients are vital for the antagonist. The involvement of b-1,3-glucanases in another hyperparasite interaction, that between Fusarium solani and Puccinia arachidis, has recently also been reported and discussed. Other elements suggesting enzymatic activity besides the dissolution of the host cell wall, are the lack of indentation of the host wall at the contact site, and the minimal swelling of the infecting hyphal tip.

Like congeneric hyperparasites, C. tenuissimum is endowed with alternative modes of penetration, since it can also invade propagules by physical destruction of the spore wall. This type of direct penetration uses a simple mechanical process of physical pressure against the host cell wall. Hyphae often coil around the rust spores, displacing the spinules as they advance, in some cases producing a swollen structure, and gaining access to the cell by breaching the spore wall, with or without the production of appressoria. A histological examination of the infection process reveals that when appressoria are produced, they generate penetration pegs that pierce the spore wall. Penetration pegs are narrower at the point where they pass through the spore wall, but once they have entered the cell lumen they swell out again. The hyperparasite destroys the protoplast of the host cells it invades and its mycelium proliferates inside the cells. Degradation of the spore contents is also evident from the many shrunk, collapsed and empty spores. While the inner wall layer of the rust spores. While the inner wall layer of the rust spores therefore remains almost intact, the cell content is probably completely digested by the combined action of the b-1,3-glucanases and other lytic enzymes. It is supposed these enzymes work in cooperation with b-1,3- glucanases, there being no chitinase production by the antagonist when grown on medium with chitin as the sole carbon source. Moreover, the great number of ungerminated spores suggests that, especially in the early stages of penetration, toxic metabolites are secreted whose role is important since they kill the host cells and thus facilitate the colonization process.

In brief, the parasitization of rust spores by the antagonist, as shown under the microscope, is divided into the following sequential events:

- Pre-penetration (signal interplay with recognition, contact, adhesion, antibiosis, formation of infection structures);

- Penetration (production of degrading enzymes, spore entry by mechanical pressure);

- Post-penetration (evasion from the host cell, sporulation).

Ecological Fitness of C. Tenuissimum

A precondition for the biological control of plant parasites is a full understanding of how the control agent operates. C. tenuissimum showed itself to be a destructive, unspecialized hyperparasite of rust fungi. Research has elucidated some of the basic principles underlying inter-fungus parasitism, clarified the fine structure of the rust–hyperparasite interface, shown that the hyperparasite inhibits rust propagules in vitro and rust diseases in planta under glasshouse conditions, and explored how the hyperparasite affects the target host. This last part of the research effort has led to the discovery of some lytic enzymes and toxic metabolites that are important pathogenicity determinants. Among the antagonist's 'weapons', these toxic metabolites are probably of prime importance. They are the cladosporols, a family of related compounds with strong antifungal activity, of which cladosporols B, C, D and E are described for the first time and reported as being produced by C. tenuissimum. The basic role of these substances in nature is to preserve the ecological niche of the hyperparasite: they protect the fungus against competing microorganisms; they prevent the growth of saprophytic microbes; and they displace plant pathogens from the plant surface. The enormous potential of such molecules to control fungal parasites is underlined by recent findings in pharmacotherapy, where two compounds (caspofungin and micafungin), having the same biological activity as the cladosporols (inhibition of the synthesis of b-1,3-glucan), were reported as a new generation of antifungal drugs. These antifungal compounds have been patented and launched on the market by pharmaceutical companies Merck and Fujizawa as the first commercial inhibitors of b-1,3-glucan synthesis, and are claimed to be effective against several fungal infections in humans.

However, the fact that C. tenuissimum is a destructive hyperparasite that attacks and disintegrates rust spores, remains viable over a wide range of temperatures, possesses several aggression mechanisms, produces fungicidal metabolites and strongly suppresses rust development in planta does not guarantee it will be effective under natural conditions. A thorough understanding of hyperparasite biology, ecology and fitness is needed to obtain effective disease control and avoid inconsistency in efficacy. The antagonist has first to survive application, then to establish itself in the environment (forest or agroecosystem), and finally it must remain active until it is required for control. This means that it has to spend a significant period of time in a permanent habitat where it has to cope with environmental constraints and live side by side with the indigenous, competing microbial community. Inability to overcome such limitations may represent a crucial bottleneck for the hyperparasite.

The occurrence of C. tenuissimum at various latitudes and altitudes indicates, however, that the physical environment does not particularly affect its survival. Similarly, nutrient availability is not a problem for this microorganism, since it has a sufficiently broad range to exploit alternate hosts. It thrives on several rust species, on plant or animal hosts, it survives saprophytically on moribund or dead plant material, and it can overwinter on dead leaves, fallen flowers or fruits, or in necrotic spots. Such versatility indicates that its life in natural habitats is quite stable, as all these substrates are permanent or semi-permanent trophic reservoirs from which the microbe can disperse into the target host at its first appearance.

Among the attributes a good hyperparasite must have to be a successful BCA are, according to Wicker and Shaw: a wide range (overlapping with that of its hosts); ecological amplitude (ensuring persistence within the host range); the production of abundant inoculum (necessary for epiphytotics to break out); an effective mode of action (to restrict the target disease); high infectivity; and virulence. An important attribute that should be added for the particular control of rust fungi is that the period of maximum sporulation of the hyperparasite should coincide with the maturation time of the rust spores. The aerobiological study of Ainsworth on the amount of Cladosporium inoculum in the air at different times of year, showed that it reached a peak in the spring and summer. These findings were confirmed by Cammack, who reported that the release of Cladosporium air-spores from senescent or dead leaves was favoured by alternating wet and dry weather, a condition that frequently occurs in temperate regions of the world as a result of the diurnal variation in late spring and early summer. The peak of C. tenuissimum sporulation therefore coincides with multiplication of rust spores (ecidio-, uredinio-, teleuto- and basidiospores) and this is further evidence that rust biocontrol with C. tenuissimum is feasible.

Authors sceptical about biocontrol usually assert that antagonists already occur in natural habitats, and yet epidemics still continue to break out. If we accept this argument, it would mean
that the whole concept of disease control was vitiated, not only that of plants but that of all
living organisms. Fortunately, however, the real successes achieved in controlling a number of
important diseases shows that those suspicions are unfounded. The upsurge of a disease over
time and space depends on a repeated cycle of infection, production of inoculum, and dispersal
of inoculum to new sites. Pathogen inoculum spread is central to the development of any disease epidemic, and in the same way antagonist inoculum spread is fundamental in achieving
control. Fluctuating climatic factors (temperature, relative humidity, precipitation, solar radiation, wind) and soil characteristics (texture, organic matter content, cation exchange capacity,
moisture, pH) strongly affect propagule persistence in the epigeal and hypogeal milieu, as well
as the dynamics of fungal populations. It is because of climatic and edaphic variations that the
distribution of diseases in stands is patchy and the occurrence of hyperparasites erratic. For this
reason the hyperparasite may need a long time to build up and maintain its biomass at levels
that will control a target pathogen. If conditions conducive to high levels of infection are not
forthcoming, the hyperparasite will be slow-acting and the pathogen will have time to cause
disease symptoms and to reproduce on the crop. Purely epidemiological factors, therefore, often
explain BCA ineffectiveness. A correct assessment of these, as well as of biological, ecological
and technical (application methods) factors, are essential for the successful exploitation of BCAs
in plant diseases.

Powdery Mildew

Powdery mildew occurs on many different flowers, woody ornamentals and trees. Several different
genera of fungi cause powdery mildew. Although usually one genus specifically attacks one or two
different plants, some species of powdery mildew (such as Golovinomyces cichoracearum formerly
Erysiphe cichoracearum) attack a wide range of plants. All the powdery mildew fungi are obligate
parasites, requiring live tissue to grow and reproduce. In greenhouses, the fungus survives by
spreading from the diseased plants to the new plants of that same crop. If that crop is not grown
for several weeks, the fungus dies out and diseased plants must be brought into the greenhouse
to establish the fungus again. Outdoors, fungal structures form on leaves and twigs that allow the
fungus to survive winter conditions.

Powdery mildew on Poinsettia

Symptoms

- White powdery fungus grows on the upper leaf surface of the lower leaves.

- Leaves may be twisted, distorted, then wilt and die.

- On some plants such as kalanchöe, infected leaves have dry, corky, scab-like spots and fungal growth is not obvious.

Conditions Favoring Powdery Mildew

- High relative humidity at night.

- Low relative humidity during day.

- 70-80 F (22-27C) temperatures (These conditions prevail in spring and fall).

The spores are carried by air currents and germinate on the leaf surface. Liquid water on leaves inhibits spore germination. The fungus grows on the leaf surface but sends fine threads (haustoria) into the cells to obtain nutrients. From the time a spore germinates to the time new spores form may require only 48 hr. High humidity favors spore formation while low humidity favors spore dispersal.

Some powdery mildew are inhibited by free moisture on leaves while others are favored by wetness on leaf surfaces.

Powdery Mildew Diseases of Ornamental Plants

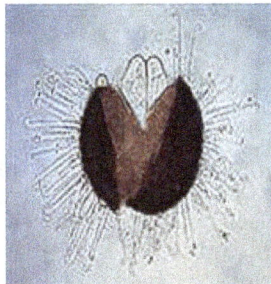

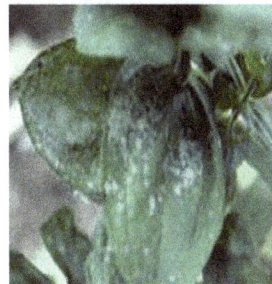

Powdery mildews are among of the most common diseases of ornamentals; many flowers, vegetables, and woody plants are susceptible. Greenhouse crops prone to infection include African violet, Begonia, Dahlia, gerbera daisy, Hydrangea, roses, Verbena, Kalanchoe, and Poinsettia.

Herbaceous perennials particularly susceptible to powdery mildew include Aster, Centaurea, Coreopsis, Delphinium, Monarda, Phlox, Rudebeckia, and Sedum.

Signs and Symptoms

The disease is easily recognizable as a white to gray powdery growth on leaves and sometimes stems and flowers. It is usually most severe on the tops of leaves, but can appear on the undersides as well. Mild cases of powdery mildew may have little or no affect on the plant other than diminishing its aesthetic value; on other instances, infected leaves may become distorted, discolored, and die prematurely. In general, powdery mildews have evolved to avoid killing their hosts because they need living plant tissue in order to survive. Symptoms and their severity depend upon the cultivar or species of host plant, the powdery mildew species, environmental conditions, and the age of plant tissue when it first became infected. Sedum develops brown scabby spots that can be mistaken for a leaf spot disease or spray injury.

Disease Cycle

Powdery mildew diseases are caused by species of fungi such as Erisyphe, Leveillula, Microsphaera, Podosphaera, Odium, and Sphaerotheca. Each powdery mildew species is specialized to infect only hosts in one genus or one family; it is rare that more than one family is affected by a single species. Infection does not spread to species of plants in other plant families. Erisyphe has a wide host range and can infect many plants in the Asteraceae family, while Sphaerotheca pannosa var. rosae is confined to roses. Podosphaera xanthii infects Calibrachoa, Verbena, and petunia, and also infects cucurbits (pumpkins, squash, melon and cucumber), so it is important to avoid growing squash and cucumber transplants in the same greenhouse as susceptible verbena or calibrachoa.

The distinctive whitish powder on leaves is composed of fine threads of fungal vegetative tissue (mycelium) and light colored mats of asexual spores (conidia). Some powdery mildews produce conidia on short, erect branches that resemble tiny chains, while others form threads so sparse that the mildew cannot be seen without the aid of a microscope. These spores are easily moved by air movement and water splash. Because powdery mildews are obligate parasites, they do not require plant stress or injury to infect plants. When spores land upon a susceptible host, they germinate and send a specialized feeding structure into the epidermis and obtain their nutrients from the plants. The infection process may take as little as 3 days or as long as 7 days. The pathogen survives in the greenhouse in weed hosts or on crops. Outdoors, the pathogen can overwinter as mycelium in infected plant parts or in resting structures (chasmothecia) produced by sexual means and visible as small, dark specks on dying leaves.

Powdery mildews, unlike most other fungal diseases, do not need free water to germinate and infect. They are favored by high relative humidity (greater than 95%), moderate temperatures (68 °-86 °F), and low light intensities. These diseases are more prevalent in the spring and fall when large differences between day and night temperatures occur. Temperatures above 90°F kill some powdery mildew fungi and spores, and the presence of free water can reduce spore germination.

Biocontrol to Kill and Prevent White Powdery Mildew

Potassium Bicarbonate

Potassium bicarbonate is a safe, effective fungicide that kills spores on contact. Like baking soda, it is also a great preventative treatment because it raises the pH level above 8.3—an alkaline environment that is not ideal for fungal growth.

- How to use:
    - Mix 3 tbsp. of potassium bicarbonate, 3 tbsp. vegetable oil, and 1/2 tsp. soap into a gallon of water. Spray onto affected plants.

Milk

Numerous studies have shown milk and/or whey to be even more effective at killing powdery mildew than chemical fungicides. In a 2009 study by the University of Connecticut, which tested a milk treatment of 40% milk and 60% water on plants infected with powdery mildew, "the milk treatment provided significantly less disease than the untreated control, and the chemical treatment had equal or significantly less disease than the milk." Scientists are not sure why milk is so effective, but they believe that when milk interacts with the sun, it produces free radicals that are toxic to the fungus.

- Ways to use:
    - Mix 60 parts water with 40 parts milk or whey, and spray onto the affected plants bi-weekly. You can even use whole milk without dilution for a strong effect.
    - Mix 1 oz. powdered milk to 2 liters of water, and spray onto affected plants bi-weekly.

Milk may be more effective at killing powdery mildew than even chemical products.

Neem Oil

Neem oil is made from the seeds and fruit of the evergreen neem tree, and it is powerful enough to kill powdery mildew in less than 24 hours. The oil works by disrupting the plant's metabolism and stopping spore production. Neem oil is also a great insecticide and since spores can be carried by bugs, this oil is a great preventative treatment as well.

- How to use:
    - Mix 3 tbsp. of neem oil to one gallon of water, and spray onto affected plants every 7-14 days. Take precautions to avoid sunburning the leaves, and avoid spraying the plant's buds and flowers.

Vinegar

The acetic acid in apple cider vinegar is very effective in killing powdery mildew. Take care to not make the mixture too strong as the acidity of the vinegar can burn plant leaves.

Mix 4 tbsp. of vinegar (5% solution) with 1 gallon of water. Reapply every three days.

Baking Soda

Baking soda has a pH of 9, which is very high. Treating with baking soda raises the pH level on the plants and creates a very alkaline environment that kills fungus. There have been mixed reports of success when using baking soda to treat severe cases, so it may be better as a preventative treatment than a fungicide.

- How to use:

  - Mix 1 tbsp. of baking soda and 1/2 tsp. liquid hand soap with one gallon of water.

  - Spray solution on affected leaves, and dispose of any remaining solution.

  - Do not apply during daylight hours. It may be best to test one or two leaves to see if the solution will cause the plant to suffer sunburn.

Baking soda's high pH creates a high alkaline environment that is unsuitable for fungi.

Garlic

Garlic has a high sulfur content and is an effective anti-fungicide. Garlic oil can be bought commercially if you do not wish to make the solution at home. It works best when added to organic oil mixtures.

- How to use:

  - Crush six cloves of garlic and add to one ounce of an organic oil such as neem oil and one ounce of rubbing alcohol. Let set for two days.

  - Strain and retain the liquid and crushed garlic.

  - Soak the garlic again (this time in one cup of water for a day). Strain out and dispose of the crushed garlic.

  - Add the oil and alcohol mixture and garlic water to one gallon of water.

  - Spray your plants, coating only the leaves.

Sulfur

Sulfur is a natural product that is very effective at preventing and controlling powdery mildew. Sulfur can be bought as a dust or as a liquid and can be added to sulfur vaporizers.

- How to use:
    - Follow the dosing instructions closely and wear gloves, eye protection, and a face mask. Avoid inhaling or coming into contact with the sulfur.

Copper Fungicides

Copper is a very effective fungicide, but it is very important to follow label directions closely. Too much copper will be detrimental to the plant and the soil.

Warning

Some ingredients, such as vinegar and baking soda, can cause sunburn to your plants. Ensure that plants are well-watered before applying and don't apply during daylight hours.

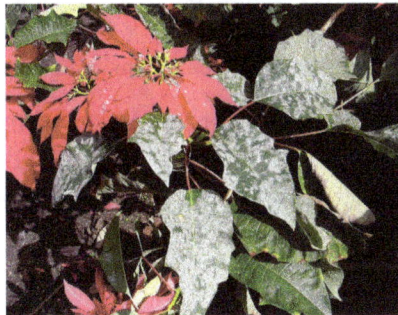

Powdery mildew on poinsettia plant.

Prevention of Powdery Mildew

Preventing the spread and/or severity of powdery mildew is the most cost-effective way of dealing with the fungus. Powdery mildew thrives in temperatures 50-65 degrees Fahrenheit with humidity levels of 80-90 percent. To prevent powdery mildew from forming in the first place, avoid low-temperature, high-humidity environments.

Do not Crowd Plants

Good air circulation ensures lower humidity levels, inhibiting the growth of powdery mildew. Crowded plants also provides too much shade for the lower leaves, which encourages fungi growth.

Do not Grow Susceptible Plants in the Shade

Powdery mildew does not tolerate high temperatures. Direct sunlight helps stem the growth of mildew because the sun's strong rays kill spores before they can spread. Plants that are shaded much of the day will stay cooler, thus encouraging the growth of mildew.

Dispose of Infected Leaves and Stems

Never use infected plant leaves or fruit as mulch or compost. Trim off infected leaves and stems and dispose of them properly. If your municipality allows backyard fires, then burn the debris. If not, dispose of the debris according to your local plant disposal regulations.

Water the Soil and not the Plants

While water itself will not encourage mildew growth, splashing the leaves with water will spread the spores. Run a hose to the base of your plants instead of using a sprinkler system.

Buy Mildew-Resistant Varieties

Powdery mildew on pumpkin leaves

There are a large variety of hybrid plants that are resistant or tolerant to the growth of powdery mildew. The resistant plants will be less likely to develop the mildew. The tolerant plants will show fewer ill-effects of an infestation of the fungi.

References

- Rust-diseases-of-ornamental-crops: umass.edu, Retrieved 5 February, 2019

- Biocontrol-of-Rust-Fungi-by-Cladosporium-tenuissimum: researchgate.net, Retrieved 24 July, 2019

- Powdery-mildew: psu.edu, Retrieved 7 March, 2019

- Powdery-mildew-diseases-of-ornamental-plants, greenhouse-floriculture: umass.edu, Retrieved 17 January, 2019

- Organic-Methods-of-Exterminating-Powdery-Mildew, gardening: dengarden.com, Retrieved 26 April, 2019

Biocontrol of Weeds

Weeds are unwanted plants in human controlled settings such as lawns, farms and fields. Weed control is a vital method in agriculture that aims to stop the growth of weeds. There are various methods employed to control weeds. This chapter discusses in detail the theories and methodologies related to the biocontrol of weeds using microbes.

Weed

A weed is any plant that requires some form of action to reduce its effect on the economy, the environment, human health and amenity. Weeds are also known as invasive plants.

Characteristics of weeds

Certain characteristics are associated with and allow the survival of weeds. Weeds posses one or more of the following:

a) Abundant seed production;

b) Rapid population establishment;

c) Seed dormancy;

d) Long-term survival of buried seed;

e) Adaptation for spread;

f) Presence of vegetative reproductive structures; and

g) Ability to occupy sites disturbed by human activities.

There are approximately 250,000 species of plants worldwide; of those, about 3% or 8000 species behave as weeds.

Weeds are troublesome in many ways. Primarily, they reduce crop yield by competing for water, light, soil nutrients, and space. Other problems associated with weeds in agriculture include:

a) Reduced crop quality by contaminating the commodity;

b) Interference with harvest;

c) Serve as hosts for crop diseases or provide shelter for insects to overwinter;

d) Limit the choice of crop rotation sequences and cultural practices; and

e) Production of chemical substances which are toxic to crop plants (allelopathy), animals, or humans.

Biological Control of Weeds

Plant invasions cause serious threat to the existence of endangered species and the integrity of ecosystems, which cost national economies tens of billions of dollars every yea. Weeds have been noted by organic horticulture producers as one of the most expensive, time consuming and troublesome activities in production. Weeds are the most significant of the economic and environmental crop loss factors and much of the weedicides applied all over the world are targeted at them. Invasive weeds cause enormous environmental damage. Also according to weeds disrupt the ecology and the functioning of rangeland plant communities and decrease the quality of services and commodities obtainable from this diverse and important natural resource. In the developing countries, weeding accounts for up to 60% of the total pre-harvest labor input and this is usually by use of simple hand tools. Weeds are generally defined as plants growing where they are not wanted. Popular methods of weed control such as mechanical and chemical are known to be: expensive, energy and labor intensive, require repeated applications, and are unsuitable for managing wide spread plant invasions in ecologically fragile conservation areas or low-value habitats, such as range lands and many aquatic systems. Also mechanical methods cause soil disturbance that may eventually lead to erosion; chemical herbicides cause environmental pollution that pose dangers to human health and wildlife, and certain weed species have developed resistance to some chemical herbicides. Biological approach to weed control dates back from 1795 when Dactylopius ceylonicus was introduced to control drooping prickly pear (Opuntia vulgaris Miller) over a large area of land; and since then biological control of weeds have been mainly through the classical strategy of introducing natural enemies from areas of co-evolution.

Biological control agents usually target their specific natural enemy weeds. Recently due to certain favorable environmental, health, economic and sustainability reasons; foreign and native organisms that attack weeds are being evaluated for use as biological control agents that may be used to complement conventional methods especially where some weeds have developed resistance to chemical control. Wheeler et al. reported that their international team discovered and tested numerous new species of potential biological control agents that could attack different plant tissues such as defoliators, sap-suckers, stem borers, and leaf- and stem-gall formers. Many successful biological weed control programs in many parts of the world have demonstrated the potency of this approach and support the concept that natural enemies can contribute to the reduction of plant growth and reproduction. Wapshere et al. classified biological approach to weed control as follows: the classical or inoculative method which is based on the introduction of host-specific exotic natural enemies adapted to exotic weeds; the inundative or augmentative method which is based on the mass production and release of native natural enemies usually against native weeds; the conservative method which is based on reducing numbers of native parasites, predators and diseases of native phytophages that feed on native plants; and the broad-spectrum method which is based on the artificial manipulation of the natural enemy population so that the level of attack on the weed is restricted to achieve the desired level of control. According to McFadyen classical method is the predominant method in weed biocontrol. He further explained that classical method involves the introduction and release of agents in form of exotic insects, mites or pathogens to give permanent control, while inundative involves the releases of predators, use of bioherbicides and other integrated pest management which usually are not as widely used as the classical method. Also there are three different techniques for applied biocontrol:

1. Conservation—protection or maintenance of existing populations of biocontrol agents;

2. Augmentation—regular action to increase populations of biocontrol agents, either by periodic releases or by environmental manipulation; and

3. Classical biocontrol—the importation and release of exotic biocontrol agents, with the expectation that the agents will become established and further releases will not be necessary.

Louda and Masters stated that despite the positive impact of chemical herbicides in agricultural productivity, complete reliance on chemical control has caused severe problems such as high cost per unit area, decreasing effectiveness, negative impact on plant diversity and increased environmental contamination. He therefore pointed out that the use of biological factors that naturally limit weed populations is one promising alternative. Menaria discussed bioherbicides as an eco-friendly approach to weed management. He explained that the use of chemical herbicides leaves some chemical residues in food commodities which directly or indirectly affect human health. According to him this situation led to the search for alternative methods that are environmentally friendly, and biocontrol has been found a suitable alternative. Green reviewed the potential for control using bioherbicides of four important forest weed species in the UK; including bracken, bramble, Japanese knotweed and rhododendron. They concluded that rhododendron is a suitable target weed for control using wood-rotting fungus as a bioherbicide stump treatment; and this is an approach already developed for weedy hardwood species in South Africa, Canada and Netherlands. Clewley et al. analyzed factors associated with control programs (invasive region, native region, plant growth form, target longevity, control agent guild, taxonomy and study duration) in order to identify patterns of control success. They found out that biological control agents significantly reduced plant size (28 ± 4%), plant mass (37 ± 4%), flower and seed production (35 ± 13 and 42 ± 9%, respectively) and target plant density (56 ± 7%).

Underlying Principles and Procedures for Biological Weed Control

Underlying Principles

The underlying principle behind biological approach to weed control is based on some research works that reported that exotic plants become invasive because they have escaped from the insect herbivores and other natural enemies that limit their multiplication and distribution in their native regions however some other factors may contribute to the tendency for particular plant species to become invasive. Therefore biological control involves using specific natural enemies that can diminish the development and reproduction of their prey organism and put some limitations to them. McFadyen stated that the predominant approach to classical biological weed control involves the importation, colonization, and establishment of exotic natural enemies (predators, parasites, and pathogens) to diminish and maintain exotic pest populations to densities that are economically insignificant.

General procedures

Some authors have outlined general procedures to be followed when embarking on classical biological weed control programs as follows:

1. Evaluate the ecology, economic impact of the weed and potential conflicts of interest;

2. Survey the organisms that are already attacking the weed in the new habitat in order to distinguish accidentally introduced agents and so eliminate such from future evaluation;

3. Carry out literature search and other forms of survey to identify natural enemies attacking the weed in its native region;

4. Screen the possible biological control agents in the foreign country to determine host range and specificity, and to remove nonspecific agents from further consideration;

5. Carry out further tests of promising candidates in quarantine after introduction to ensure host specificity and eliminate predators, parasites, and pathogens that may have been introduced with them;

6. Embark on mass rearing of host-specific agents;

7. Release the host-specific agents;

8. Carry out post-release evaluation to determine establishment and effectiveness of agents; and

9. Redistribute agents to other areas where control is required. Wapshere et al. presented a summary of steps normally followed when introducing a biological control agent in a classical biological control weed program as in table below:

Steps	Details
1. Initiation	Data on taxonomy, biology, ecology, economics, native and introduced distributions, known natural enemies, etc., are compiled by initiating scientist or group. An extensive literature review is conducted on the proposed target weed and its relatives, plus known natural enemies. Conflicts of interest identified and resolved if possible
2. Target weed approval	Data in (step 1) submitted to appropriate State and Federal groups for comment; additional data may be required
3. Foreign exploration and domestic surveys	If project approved in (step 2), the center of evolution of the genus of the target weed (if known) and other suitable areas, are searched for natural enemies, particularly where these are eco-climatically similar to the area of introduction. At the same time, the weed should be investigated in the country of introduction for attacking enemies, related plants, etc.
4. Weed ecology and agent host specificity	Ecology of the target weed, its close relatives and its natural enemies is studied in the native area, and the most damaging and apparently selective agents are subjected to several years of host-specificity testing
5. Agent approval	A report on each agent is submitted to appropriate State and Federal bodies to obtain importation and release permits
6. Importation and quarantine clearance	Each agent is imported to the country of introduction where it is reared through at least one generation in quarantine to rid it of its parasites and diseases

7. Rearing and release	After a pure culture of the agent is obtained in (step 6), it is normally mass-reared and released in the field in cages or free at field sites
8. Evaluation and monitoring	Agent is monitored at field sites to determine establishment and degree of stress on target weed, or to determine reasons why the agent did not become established or efficacious
9. Redistribution	To aid spontaneous self-dissemination, agent is distributed to other areas in the target weed's distribution, if needed

Reasons for Relatively Slow Popularity and Adoption of Biological Weed Control

Recent research activities and weed control practices around the world have shown that the old idea derived from untested opinions; that biological approach to weed control is usually very slow, unpredictable, expensive and mostly unsuccessful is totally not true. Apart from the high initial costs, biological approach to weed control has been known to be relatively cheaper when compared to other methods; however certain factors have slowed down the rate of adoption. These factors include: long time of establishment-usually 20 years or more to ensure success, inadequate or no records of the extent of pre-biological control weed infestations that should serve as a guide for a new biocontrol program, discouraging story of poorly implemented weed bio-control programs. A lot of success stories however have been documented. Lack of information about previous successfully implemented biological control of weeds often lead to untested theories becoming established dogma and this negatively influence the decisions to or not apply it. For instance Mcfadyen stated that it was believed that biological control of trees is difficult, but many examples of trees controlled by insects have been reported. Also classical biological control has been viewed as unsuitable for weeds of annual crops or other frequently disturbed environments, however there are many examples of successful control of crop weeds.

Some researchers have reported that there are evidences showing that some agents introduced for exotic weed control have attacked non target, native plants and this situation has raised concerns among biological control workers and weed scientists as well as the governments. Opposition to biological approach to control of weeds has also contributed to slowing down the rate of adoption and practice; this is because some researchers and weed control scientists believe that it is difficult to estimate the cost or the feasibility of biocontrol. Based on a study carried out in South Africa, it was reported that some of the weed biocontrol projects have provided practical solutions to problems e.g. the development of Stumpout for the treatment of wattle stumps and the use of C. gloeosporioides for the control of H. sericea. However other projects have been less successful and have resulted in the rejection of potential agents for various reasons and these include C. albofundus on A. mearnsii, X. campestris on M. aquaticum and G. nitens on R. cuneifolius. Vurro and Evans identified legislative hurdles, technological and commercial constraints as limitations to the adoption of biological weed control in Europe. Olckers stated that limited budgets in many countries have also helped to slow than the rate of adoption and practice of biological approach to weed control.

Examples of successful biological control of weeds with introduced insects and pathogens:

One thousand one hundred and forty-four individuals (mostly entomologists and plant pathologists) have ever attended the International Symposia on Biological Control of Weeds (ISBCWs); and out of these, 450–550 weed biological control experts have been actively involved in research and development efforts over the last 50 years mainly from USA, Canada, Australia, South Africa and New Zealand. McFadyen reported that biological approach to weed control has a long history and a good success rate of 94. A comprehensive list of agents and their target weeds have been documented by Winston et al. Culliney presented potential benefits estimated for some proposed or initiated biological control programs targeting invasive weeds. Frequently cited examples of successful approach to biological weed control are the prickly pear cacti (Opuntia; spp.) in Australia, eradicated by an imported moth (Cactoblastis cactorum) and rangeland in California, Oregon, Washington, and British Columbia controlled by St. John's wort Hypericum perforatum (millepertuis perforé). Mcfadyen presented a list of 41 weds which have successfully been controlled using introduced insects and pathogens and another three weeds also controlled by introduced fungi applied as mycoherbicides. He further stated that many of these successes have been repeated in other countries and continents. Julien presented a list of both successful and failed cases of biological weed control; this included the introduction of 225 organisms against 111 weed species, and 178 insects and 6 mites. Palmer et al. reported that 43 new arthropod or pathogen agents were released in 19 projects; and that effective biological control was achieved in several projects with the outstanding successes being the control of rubber vine, Cryptostegia grandiflora, and bridal creeper, Asparagus asparagoides.

Success of Weed Biological Control

Information collated on weed impacts before the initiation of a biological control program is necessary to provide baseline data and devise performance criteria with which the program can subsequently be evaluated. For avoidance of confusion on when a biological control could be viewed as successful or not, Hoffmann stated that an implementation of a particular biological control will be termed successful when: complete-when no other control method is required or used, at least in areas where the agent(s) is established; substantial-where other methods are needed but the effort required is reduced (e.g. less herbicide or less frequent application); and negligible-where despite damage inflicted by agents, control of the weed is still dependent on other control measures. Complete control does not imply total eradication of the weed; rather it means that control measures are not required anymore specifically against the target weed, and that crop or pasture yield losses will not be attributed mainly to this weed. Substantial control involves situations where control may be complete in some seasons and/or over part of the weed's range, as well as cases where the control achieved is widespread and economically significant but the weed is still a major problem. It is therefore concluded that successful implementation of biological approach to weed control is the successful control of the weed, and not necessarily the successful establishment of individual agents released against the weed. Successful biological control depends on three factors: the extent to which each individual agent can limit the targeted plant; the ecology of the agent as it affects its ability to populate and spread easily in the new environment; and the ecology of the weed, which determines if the total damage that can be caused by the agent can significantly reduce its population. Because agents always need some surviving predator plants to complete their life cycle, biological control will not usually totally eradicate their target weeds. In essence a successful biological control program reduces the potency and population of the target weed and usually in conjunction with other control methods as part of an overall integrated weed management scheme which is recommended.

Things to Consider when Making the Choice of Agents to be Introduced to Control Weeds:

Gassmann reported that selection of potential agents in the last decades has been mainly based on the population biology of the weed, impact studies of agents on the plant and the combined effect of herbivory and plant competition. Palmer et al. stated that agent selection is highly dependent on the type of weed, its reproductive system, on the ecological, abiotic and management context in which that weed occurs, and on the acceptable goals and impact thresholds required of a biological control program.

Generally, factors to be considered in selecting agents include the following: the agent must target a particular plant species, must have high level of predation and parasitism on the host plant and its entire population, must be prolific, must be able to thrive in all habitats and climates where the weed exists and should be able to spread easily and widely, must be a strong colonizer, the overall cost of introducing the agent must be cheaper compared to other control methods, the technology that will be involved in introducing and managing the agent must be as simple as possible, must as much as possible maintain natural biodiversity, sufficient number of individuals must be released, plant phenology (effect of periodic plant life cycle events) must be favorable. To be considered a good candidate for biological control, a weed should be non-native, present in numbers and densities greater than in its native range and numerous enough to cause environmental or economic damage, the weed should also be present over a broad geographic range, have few or no redeeming or beneficial qualities, have taxonomic characteristics sufficiently distinct from those of economically important and native plant. Furthermore, the weed should occur in relatively undisturbed areas to allow for the establishment of biological control agents, cultivation, mowing and other disturbances can have a destructive effect on many arthropod biocontrol agents. Inundative biocontrol agents such as bacteria and fungi are less sensitive to these types of disturbances so may be used in cropland.

Steps to Identifying and Introducing Biological Control Agents:

The study of insect attributes and fitness traits, the influence of plant resources on insect performance, and the construction of comparative life-tables, are the first steps towards an improvement of the success rate of biological weed control. Generally, steps to identifying and introducing biological control agents include:

1. Identify control agents and determine the level of specialization;

2. Identify target weeds;

3. Apply controlled release of the agents;

4. Apply full release and determine optimal release sites;

5. For the case of classical methods, monitor release sites;

6. Apply redistribution for the case of classical methods;

7. And maintain control agent populations.

Biocontrol of Weeds with Microbes

Biological Control of Weeds using Fungi

Most commercial biological weed control products researched in North America have been based on formulations of fungal species, however, few have been successful in the long term. Examples include BioMal, a formulation of Colletotrichum gloeosporioides f.sp. malvae, introduced for the control of round leaf mallow (Malva pusilla), and C. gloeosporioides f.sp. aeschynomene, which was released for control of northern jointvetch (Aeschynomene virginica) in the United States in 1982 as Collego, and again in 2006 as LockDown. Additionally, Sarritor, a formulation of Sclerotinia minor was introduced for the control of dandelion (Taraxacum officinale), white clover (Trifolium repens) and broadleaf plantain (Plantago major) in turf.

Within the scientific literature, three genera of fungi have received the majority of attention as bioherbicide candidates. In addition to the aforementioned BioMal and Collego, several other species within the genus Colletotrichum have been investigated. Additional examples include C. truncatum, which has been investigated to control hemp sesbania (Sesbania exaltata), and C. orbiculare, which was investigated for its potential to control spiny cocklebur (Xanthium spinosum). An investigation of the genomes of C. gloeosporioides and C. orbiculare, found that both species contained a number of candidate genes predicted to be associated with pathogenesis, including plant cell wall degrading enzymes and secreted disease effectors including small secreted proteins (SSPs), the latter of which were shown to be differentially expressed in planta according to stage of infection, suggesting that some of these proteins may have specific roles in the infection process. There is also evidence that both of these Colletotrichum species have the ability to produce indole acetic acid, a plant hormone, derivatives of which are well established herbicide templates.

Three species within the genus Phoma have also received attention as potential agents for biological weed control. P. herbarum, a fungal pathogen originally isolated from dandelion leaf lesions in Southern Ontario, has been investigated for control of dandelions in turf. P. macrostoma has also been investigated for similar purposes as it has been observed to specifically inhibit the growth of dicot plants. The 94-44B strain of this species has been registered for control of broadleaf weeds in turf systems in Canada and the US. An investigation of 64 strains of P. macrostoma, including 94-44B, found that the bioherbicidal activity of these species was limited to a genetically-homogeneous group of strains, all of which were isolated from Canada thistle. Through mass spectrometry, P. macrostoma has been recognized to produce photobleaching macrocidins that do not affect monocots. As the activity of P. macrostoma is most apparent on new growth, it has been suggested that these compounds are transported in the phloem of the host plant. Unfortunately, the specific phytotoxic mechanism of macrocidins remains unknown. Despite this, macrocidins and other molecules within the tetramic acid family have received significant attention as templates for the development of novel synthetic herbicides. Additionally, an anthraquinone pigment has been isolated from a P. macrostoma strain and shown to have herbicidal effects on several prominent weeds of Central India. Anthraquinone pigments produced by other fungi have also been demonstrated to cause necrosis on wheat leaf blades and a variety of cultivated legumes. Although the phytotoxic mechanism underlying the effects of these compounds has not been fully characterized, the development of necrosis after exposure to the anthraquinone lentisone was found to be light dependent, a potential clue for the eventual determination of the mechanism associated with this class of molecules. Also of note within this genus is Phoma chenopodicola, which has been investigated as a potential control agent for lamb's quarters (Chenopodium album). A phytotoxic diterpene, chenopodolin, has been isolated from this species, which was found to cause necrotic lesions on lamb's quarters (Chenopodium album), creeping thistle (Cirsium arvense), green foxtail (Setaria viridis) and annual mercury (Mercurialis annua). Two additional fungal isolates of the genus Phoma have also been found to cause a modest degree of stem rot on C. arvense, however, these isolates were not identified at the species level.

Two species within the aforementioned Sclerotinia genus have been investigated for their potential to control weeds. Abu-Dieyeh and Watson found that Sclerotinia minor effectively controlled dandelions with and without the presence of turf species in greenhouse conditions. A follow up trial including application of S. minor in field conditions confirmed these results. As noted earlier, S. minor strain IMI 344141 was introduced to the Canadian lawn care industry under the product name Sarritor in 2010, however, it is no longer commercially available. A relative of S. minor, S. sclerotiorum, has been observed to have phytotoxic activity against creeping thistle (Cirsium arvense). Production of oxalic acid by both S. minor and S. sclerotiorum has been observed to play a role in the virulence of these fungi on their host plant. Oxalic acid production can be encouraged through addition of sodium succinate to S. minor growth media, and cultures grown on sodium succinate-enriched media caused greater development of necrotic tissue when applied to dandelion than cultures grown on non-enriched media. Oxalic acid acidifies the host tissue, enabling cell wall degradation, and also interferes with polyphenol oxidase (PPO), which normally assists in plant defense. Low concentrations of oxalic acid have also been shown to suppress the release of hydrogen peroxide, another plant defense molecule, in cell cultures of soy and tobacco.

In addition to the other examples described earlier in the text, several other fungi have been registered as bioherbicides for use in forestry or ecosystem management in Canada and the US, though in general, there appears to be limited research about these strains with respect to their mode of action. Two separate strains of Chondrostereum purpureum have been registered in Canada and the US for controlling regrowth of deciduous tree species in coniferous plantations. This fungal species is a naturally occurring pathogen of deciduous trees in North America. Although the potential host range of this species is fairly wide, wound infection is a key element of successful infection in most cases. C. purpureum strain HQ1 was registered under the product name Mycotech Paste with the PMRA in 2002 and the EPA in 2005. Registration of this strain with the PMRA ended in 2008. Another strain of this species, PFC 2139, was registered under the product name Chontrol Paste with the EPA in 2004 and with the PMRA in 2007. Both registrations are currently active and this product remains commercially available.

Another fungus, Puccinia thlaspeos, was registered with the EPA in 2002 under the product name Woad Warrior for control of Dyer's woad (Isatis tinctoria) . This fungus is an obligate parasite and requires a living host to reproduce, however, inoculum can be produced from dried and ground plant material of its target weed. This product is no longer commercially available.

Alternaria destruens strain 059 was registered with the EPA in 2005 under the product names Smolder WP and Smolder G. This product, originally isolated from Cuscuta gronovii growing in unmanaged conditions in Wisconsin, is intended for control of dodder species (Cuscuta spp.), however, it is not commercially available.

A final bioherbicide that bears mentioning is DeVine, a formulation of the fungus Phytophthora palmivora. This product was registered with the EPA in 1981 and again in 2006. P. palmivora was originally isolated from strangler vine (Morrenia odorata) in Florida and was used to control the same species in citrus orchards. Although this product was re-registered in 2006, it is no longer commercially available.

Biological Control of Weeds using Bacteria

A number of bacteria have also been investigated as potential biological weed control agents. Of these, Pseudomonas fluorescens and Xanthomonas campestris have attracted the most attention. Biological weed control using bacteria has been suggested to have several advantages over the use of fungi, including more rapid growth of the bioherbicide agents, relatively simple propagation requirements, and high suitability for genetic modification through either mutagenesis or gene transfer.

P. fluorescens has received much of the attention as a biological weed control agent. There are many strains of this species, some of which are beneficial to plants, whereas others are inhibitory. Among studies into the suppressive effects of P. fluorescens, three strains have been investigated in especially great detail, all of which have been observed to inhibit plant growth and/or germination through the production of extracellular metabolites.

Pseudomonas fluorescens strain D7, originally isolated from the rhizospheres of winter wheat (Triticum aestivum) and downy brome (Bromus tectorum) in Western Canada, has been observed to selectively inhibit growth and germination of a number of grassy weeds, most notably downy brome. By selective removal of compounds from cell-free filtrates, the growth-inhibiting activity associated with this strain was attributed to a combination of extracellular peptides and a lipo-polysaccharide, which were suggested to work in conjunction to express herbicidal activity. No subsequent reports regarding mechanism were found in the available literature.

Conversely, P. fluorescens strain WH6 has been observed to affect the germination of a much broader range of plant species, significantly inhibiting germination of all species tested (21 mono-cot species and 8 dicot species) with the exception of a modern corn (Zea mays) hybrid. The germination-inhibiting activity of the WH6 strain has been attributed to the production of a compound originally referred to as Germination Arrest Factor. The active component of GAF has been identified through nuclear magnetic resonance spectroscopy and mass spectrometry as 4-formylaminooxy-L-vinylglycine, and its biosynthesis has been proposed to begin with the amino acid homoserine. This class of compounds, the oxyvinylglycines, has been shown to interfere with enzymes that utilize pyridoxal phosphate as a cofactor, including enzymes involved in nitrogen metabolism and biosynthesis of the plant hormone ethylene. Interestingly, GAF has also been recognized to express specific bactericidal activity against Erwinia amylovora, the bacterium that causes fire blight in orchards. The genome sequence of P. fluorescens strain WH6 has been published, and gene knockouts were used to identify several biosynthetic and regulatory genes involved in GAF/4-formylaminooxy-L-vinylglycine production. Strain D7 was also included in the original investigation of strain WH6, however, as culture filtrates of strain D7 prepared in the same manner as WH6 did not possess germination-inhibiting activity the authors suggested that GAF was not responsible for the activity associated with strain D7.

The production of extracellular metabolites with phytotoxic effects has also been observed in an additional P. fluorescens strain, referred to as BRG100, which has been recognized to have suppressive activity on the grassy weed green foxtail (Setaria viridis). The herbicidal compounds produced by this species, referred to as pseudophomin A and B, have been characterized through a combination of serial chromatography, high performance liquid chromatography (HPLC), thin layer chromatography (TLC), chemical degradation, and X-ray crystallography. Unfortunately, neither the biosynthetic pathway involved in the production of these compounds nor the specific biochemical effects of these molecules on green foxtail have been characterized at this time. However, the full genome sequence of this strain has been published and a detailed projection of the costs and technical requirements for the mass production of this biocontrol agent has been reported.

The other bacterial species that has received much of the attention as a candidate biological weed control agent is Xanthomonas campestris. Most notably within this species, the strain X. campestris pv. poae (JT-P482) was registered in Japan in 1997 for control of annual bluegrass (Poa annua) under the product name Camperico. The activity of this species is specific to Poa annua and Poa attenuata, and was not reported to affect other turf species tested. A separate strain of X. campestris has also received attention as a potential control agent against horseweed (Conyza canadensis). No discovery of phytotoxic compounds was reported in any of the aforementioned investigations into application of X. campestris as a bioherbicide, however, compounds with phytotoxic activity have been previously isolated from the vitians pathovar of this species, and it is possible that phytotoxic metabolites play a role in the suppression of Poa annua and Conyza canadensis. Although the cause of host-plant suppression was not indicated in the above studies, the infection process of X. campestris pv. campestris (Xcc) in brassica crops has been well characterized. Briefly, Xcc can colonize the xylem of the host plant and use this pathway to spread throughout the organism. The success of Xcc in reaching the host xylem is contingent on its interaction with receptor proteins of the host plant that can recognize pathogen associated molecular patterns (PAMPs), potentially resulting in elicitation of plant defense responses such as programmed cell death and increased production of reactive oxygen.

Biological Control of Weeds using Viruses

In select cases, viruses that affect weed species have also been considered as bioherbicide candidates. This strategy is more commonly considered for management of invasive species in broader ecosystems rather than specifically managed areas. Viruses have been suggested to be inappropriate candidates for inundative biological control due to their genetic variability and lack of host specificity. Examples of viruses that have been investigated for the potential to control invasive or undesirable species include Tobacco Mild Green Mosaic Tobamovirus for control of tropical soda apple (Solanum viarum) in Florida, and Araujia Mosaic Virus for control of moth plant (Araujia hortorum) in New Zealand. A patent on the former biological control agent has been filed and EPA approval for use on fenced-in pasture areas was granted in 2015. A virus resembling Tobacco Rattle Virus has also been proposed as a control agent for Impatiens glandulifera, an invasive weed of concern in central and western Europe. Similarly, Óbuda Pepper Virus (ObPV) and Pepino Mosaic Virus (PepMV) have been proposed as agents to reduce overall populations of the weed Solanum nigrum. The biological activities of viruses are very distinct from pathogenesis caused by bacteria or fungi, and may present additional opportunities for biological weed control in some situations.

Real World Factors that affect the Efficacy of Bioherbicides

The research pipeline from the screening stage to field conditions faces a number of unique challenges. One commonly reported challenge is the need for continuous moisture availability during the period in which the biocontrol agent infects the plant. It was reported that dew periods of more than 12 hours are commonly necessary for bioherbicide candidates to successfully infect their hosts. A variety of techniques to provide this moisture have been tested, with varying degrees of success. In order to prolong the period of leaf wetness necessary for successful infection of dandelion by Phoma herbarum, several vegetable oil emulsions were included in aqueous inoculants, however, these additives were found to be phytotoxic, thus obscuring the benefit to infection that may have been caused by their addition. Timing inoculant application to prolong the leaf wetness period (e.g., application at dawn or dusk) has also been suggested as a simple method of maximizing infection, although the success of this technique can be highly sensitive to environmental fluctuations. In some cases, solid inoculant media have also been investigated. The most common method for developing solid inoculant media is to propagate the candidate biological weed control species on grains which will subsequently be applied directly to the field or incorporated with other moisture-retaining materials such as calcium alginate, oils or vermiculite. Granular applications have the advantage of prolonging the in-field survival of introduced biological weed control agents through the provision of moisture and nutrients, however, they are also generally associated with a more gradual rate of infection.

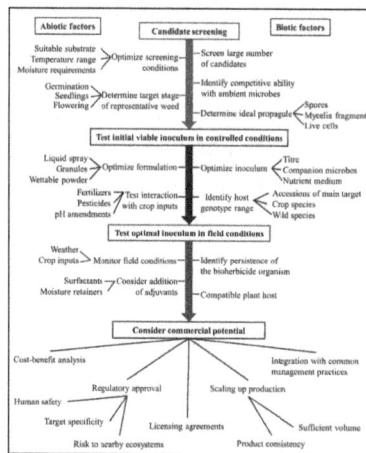

Overview of factors involved in the development of a bioherbicide.

The interplay of temperature and humidity also has a significant effect on the success or failure of infection by many pathogens and may alter the efficacy of biological control agents Figure Cold air can retain less total moisture than warm air, and thus the relative humidity is more commonly elevated at lower temperatures. Elevated humidity is generally beneficial to successful bioherbicide colonization because it decreases evaporation rates, thus increasing the duration of leaf wetness following inoculant application. In investigating the efficacy of the biological weed control product Sarritor, it was found that infection rates were highest when the temperature remained below 20 °C and the relative humidity was high. Similar requirements have been suggested by, who observed that successful infection of spiny cocklebur (Xanthium spinosum) by biological weed control agent Colletotrichum orbiculare may have been contingent on elevated humidity. As with any species, most biological control species will have a fairly finite temperature range in which they can survive, as well as a narrower range in which their activity will be maximized. For example, found that ambient temperatures below 25 °C (day) and 20 °C (night) caused decreased efficacy of Xanthomonas campestris pv. poae in the suppression of annual bluegrass (Poa annua). Similarly, the efficacy of X. campestris isolate LVA-987 in controlling horseweed (Conyza canadensis) was found to require ambient temperatures between 20 °C and 35 °C, with peak efficacy between 25 °C and 35 °C. This parameter will be different for any given biological control candidate species and thus temperature and humidity should be tracked throughout any efficacy trials involving biological weed control agents.

Quorum sensing refers to the ability of a bacterium to differentially express genes based on its population density. The effect of bacterial and fungal population densities can in some cases inform the behavior of these species, and in some cases affect whether a pathogen is virulent or latent. This is an important factor in the characterization of potential bioherbicides, however, testing inoculant media with varying population densities is not a common practice within the investigation of biological weed control strategies, nor is the phenomenon of quorum sensing commonly discussed in the related literature. However, apparent latent periods in the life cycle of biological weed control candidate species have been occasionally reported, and it is possible that quorum sensing effects could explain these cases of asymptomatic infection.

It is possible that interactions with fertilizers and pesticides could affect the infectiousness of a candidate biological weed control agent. For example, an investigation of the ability of P. macrostoma to control dandelions in turf found that co-application with a high rate of nitrogen fertilizer improved its efficacy, whereas co-application with phosphorus had no effect, and potassium sulfate decreased efficacy.

Challenges in Commercialization

Despite the promise shown by many bioherbicides, few have achieved long-term commercial success, in part due to the challenges to achieving consistent efficacy in field conditions noted above. For example, amongst the fungal bioherbicides described in this review, only LockDown (C. gloeosporioides f. sp. aeschynomene) remains commercially available. In the case of BioMal (C. gloeosporioidesf.sp. malvae), the narrow target specificity (only round leaf mallow) of the product made for a market niche that was too small to cover production costs. Additionally, significant challenges were encountered in maintaining product consistency while scaling up production volumes. For these reasons, Philom Bios, the original commercial producer of this bioherbicide, discontinued its production in 1994, only 2 years after registration. The strain was later licensed to Encore Technologies in 1998, however, challenges with maintaining product consistency under mass production led to the abandonment of the project. In the case of Sarritor (S. minor strain IMI 344141), the commercial failure has been attributed to challenges with increasing production volume and product consistency, as well as inconsistent efficacy of the product due to the narrow range of environmental conditions in which successful infection will occur.

Future Directions

Mechanism of Action

The mechanism(s) behind the suppressive activity of a given biocontrol agent is in many cases only partially understood. Future research into the mechanisms underlying these effects will be important to achieve consistent efficacy with biocontrol agents, as well as to evaluate potential impacts on human and ecosystem health. This in turn will be of value to gaining regulatory approval. Additionally, understanding bioherbicidal mechanisms may generate novel chemical herbicides to overcome current resistance traits, and will likely also be of peripheral value to the field of plant pathology.

Transition to the Field

Translating effects observed in a controlled environment to field conditions is a significant challenge to the development of successful biocontrol agents and it is common for projects to conclude at this juncture. Thus, the development of new delivery formulations intended to improve the in-field stability of biocontrol agents is as important as the discovery of the agents themselves. Widespread testing of a given biocontrol agent in a variety of locations, similar to plant variety testing, is essential to understanding the feasibility of introducing that agent on a broad scale. Finally, the production of commercially relevant quantities of viable inoculum or culture extract must also be considered, as techniques employed in the laboratory are frequently impractical for industrial-scale production. Lessons from other industries such as pharmaceuticals and probiotic foods will likely be valuable in addressing this challenge.

Extraction of Herbicidal Compounds

In some cases a particular herbicidal compound can be extracted from a live culture. This strategy can yield a more stable control agent, the efficacy of which will not be contingent on the continued survival of a given organism in an uncontrolled environment. Although the differentiation between naturally and synthetically sourced pesticides may be arbitrary in terms of their potential effect on human and environmental health, such compounds will likely be more acceptable in the public eye than those produced through traditional chemistry.

New Sources of Bioherbicide Candidates

This review has focused on a limited number of genera which have received an especially great degree of attention as bioherbicide candidates for turf and field crop situations. Considering the degree of taxonomic diversity among microbes, there are opportunities to employ other genera as bioherbicides in the future. However, there are additional ecological niches from which potential biological weed control candidates can be discovered. For example, most plant species form relationships with a variety of microbes, referred to as endophytes, which colonize the internal environment of the plant without causing disease. There is evidence that endophytes can play a role in nutrient accumulation, drought tolerance and disease resistance. Growth-promoting endophytes have been shown to reduce weed populations in pastures by inoculating the desired grass species, enabling them to compete with weeds more effectively. It has been reported that some plant-inhabiting microbes will express host-specific behavior, acting as an endophyte in some plant species but as a pathogen in another. Additionally, some endophytes have also been reported to produce compounds that are phytotoxic to non-host species. These phenomena could potentially be applied to controlling undesirable weed species. Endophyte-based weed control may have unique advantages over the application of pathogens such as improved ability of candidate microbes to persist in field conditions through having a more consistent ecological niche within their plant host, or the provision of other benefits to their host such as nutrient acquisition or disease resistance.

Permissions

Index